Laboratory Experiments in Microbiology

Brief Edition
Third Edition

Of Related Interest from the Benjamin/Cummings Series in the Life Sciences

GENERAL BIOLOGY

N. A. Campbell
Biology, Second Edition (1990)

PLANT BIOLOGY

M. G. Barbour, J. H. Burk, and W. D. Pitts
Terrestrial Plant Ecology, Second Edition (1987)

J. Mauseth
Plant Anatomy (1988)

L. Taiz and E. Zeiger
Plant Physiology (1991)

BIOCHEMISTRY AND CELL BIOLOGY

W. M. Becker and D. W. Deamer
The World of the Cell, Second Edition (1991)

C. Mathews and K. H. van Holde
Biochemistry (1990)

G. L. Sackheim
Chemistry for Biology Students, Fourth Edition (1990)

MOLECULAR BIOLOGY AND GENETICS

E. J. Ayala and J. A. Kieger, Jr.
Modern Genetics, Second Edition (1984)

L. E. Hood, I. L. Weissman, W. B. Wood, and J. H. Wilson
Immunology, Second Edition (1984)

R. Schief
Genetics and Molecular Biology (1986)

J. D. Watson, N. H. Hopkins, J. W. Roberts, J. A. Steitz, and A. M. Weiner
Molecular Biology of the Gene, Fourth Edition (1987)

G. Zubay
Genetics (1987)

MICROBIOLOGY

E. Alcamo
Fundamentals of Microbiology, Third Edition (1991)

R. M. Atlas and R. Bartha
Microbial Ecology: Fundamentals and Applications
Second Edition (1987)

J. Cappuccino and N. Sherman
Microbiology: A Laboratory Manual, Third Edition (1992)

G. Tortora, B. Funke, and C. Case
Microbiology: An Introduction, Fourth Edition (1992)

EVOLUTION, ECOLOGY, AND BEHAVIOR

D. D. Chiras
Environmental Science, Third Edition (1991)

R. J. Lederer
Ecology and Field Biology (1984)

M. Lerman
Marine Biology: Environment, Diversity, and Ecology (1986)

D. McFarland
Animal Behavior (1985)

E. Minkoff
Evolutionary Biology (1983)

R. Trivers
Social Evolution (1985)

ANIMAL BIOLOGY

E. N. Marieb
Essentials of Human Anatomy and Physiology
Third Edition (1991)

E. N. Marieb
Human Anatomy and Physiology, Second Edition (1992)

E. N. Marieb and J. Mallatt
Human Anatomy (1992)

L. G. Mitchell, J. A. Mutchmor, W. D. Dolphin
Zoology (1988)

A. P. Spence
Basic Human Anatomy, Third Edition (1991)

Laboratory Experiments in Microbiology

Brief Edition
Third Edition

Ted R. Johnson
St. Olaf College

Christine L. Case
Skyline College

The Benjamin/Cummings Publishing Company, Inc.

Redwood City, California • Menlo Park, California • Reading, Massachusetts
New York • Don Mills, Ontario • Wokingham, U.K. • Amsterdam • Bonn
Sydney • Singapore • Tokyo • Madrid • San Juan

Sponsoring Editor: Edith Beard Brady
Editorial Assistant: Sissy Lemon
Production Service: Larry Olsen
Copy Editor: Betsy Dilernia
Artists: Ken Miller and Cathleen Jackson Miller
Compositor: Thompson Type
Cover Designer: Mark Ong
Cover Photograph: © Will and Deni McIntyre/Allstock

ISBN 0-8053-8488-X

4 5 6 7 8 9 10–BA–96 95 94 93

Library of Congress Cataloging-in-Publication Data
Johnson, Ted R., 1946–
      Laboratory experiments in microbiology / Ted R.
Johnson, Christine L. Case. — 3rd ed., Brief ed.
          p.   cm.
      Includes index.
      ISBN 0-8053-8488-X
      1. Microbiology—Laboratory manuals.   I. Case,
Christine L., 1948–   II. Title.
QR63.J65   1992
576′.078 — dc20   92-3560
                                                   CIP

The Benjamin/Cummings Publishing Company, Inc.
390 Bridge Parkway
Redwood City, California 94065

Figure Acknowledgments

Part 1, p. 6: Biological Photo Service. Figure 1.2: AO Scientific
Instruments. Photos p. 22 a and b: Christine Case. Figure 18.1:
Roche Diagnostics, Division of Hoffman-La Roche, Inc., Nut-
ley, NJ. Part 10: B. D. Davis et al., *Microbiology,* 3rd ed., Hag-
erstown, MD, Harper & Row, 1980. Part 11, p. 250: J. G. Hadley,
Battelle-Pacific Northwest Laboratories/Biological Photo Ser-
vice. Figure 43.1: Christine Case. Part 12: G. T. Cole, University
of Texas, Austin/Biological Photo Service. Part 12, p. 276:
Dwayne C. Savage, University of Illinois at Urbana-
Champaign. Figure 51.1: T. D. Brock and K. M. Brock, *Basic
Microbiology with Applications,* 2nd ed., Englewood Cliffs,
NJ, Prentice-Hall, 1978.

Preface

Laboratory Experiments in Microbiology: Brief Edition, Third Edition is a laboratory companion to Microbiology: An Introduction, Fourth Edition by Tortora, Funke, and Case. This edition includes new exercises that reflect recent developments and current techniques in environmental microbiology and genetic engineering. Our goal is to provide a manual of basic microbiological techniques with applications for undergraduate students in diverse areas, including the biological sciences, the allied health sciences, agriculture, environmental science, nutrition, pharmacy, and various preprofessional programs. This manual contains 56 thoroughly class-tested exercises covering every area of microbiology. Most exercises require about one and one-half hours of laboratory time. By selecting an appropriate combination of exercises, the instructor can provide learning experiences to meet the needs of a particular course.

Our Approach

Our goal in writing this manual has been twofold—to teach microbiological techniques and to show students the importance of microbes in our daily lives and their central roles in nature. All of the laboratory techniques recommended by the Committee on Undergraduate and Graduate Education of the American Society for Microbiology are included in this manual.

Laboratory safety is our primary concern. Many students are preparing for work in a clinical environment and need to learn to handle biologically contaminated materials. Students preparing for work in biotechnology and other laboratories must master aseptic techniques to avoid contamination of their samples. Students must not only learn but also practice these safety techniques so that safety becomes part of their professional behavior. We have followed the Centers for Disease Control guidelines for safe handling of microbes and human body fluids. Students are instructed to work only with their own body fluids. See the section entitled Safety in the Laboratory on page 2 and the following sections of the Introduction for specific safety suggestions. At key points in the exercises appear safety boxes, marked with either a biohazard logo ☣ or a general safety logo ⚠ , indicating appropriate safety techniques.

Almost every exercise includes an actual experiment requiring students to analyze data. We hope in this way to promote analytical reasoning and to make laboratory sessions interesting for students as well as to provide a variety of opportunities for reinforcement of the technical skills they have learned. To demonstrate further some practical uses of microbiology, we have frequently included material with direct application to procedures performed in clinical and commercial laboratories.

Scope and Sequence

This manual is divided into 13 parts. Each part introduction explains the unifying theme for that part. Exercises in the first three parts provide sequential development of fundamental techniques. The remaining exercises are as independent as possible to allow the instructor to select the most desirable sequence. A Techniques Required section preceding each experiment lists prerequisites from earlier exercises.

Part 1, Microscopy, emphasizes observation through the microscope. Practice in use and care of the microscope is followed by observations of microbes, to familiarize students with their sizes.

Part 2, Staining Methods, begins with an explanation of how to handle bacterial cultures and teaches the use of the most common stains, including preparation of stained samples and their examination. A highly reliable capsule stain is included. Exercise 8, Morphologic Unknown, introduces the concept of unknowns and can be used to test knowledge of staining methods.

Part 3, Cultivation of Bacteria, stresses aseptic technique and covers the isolation of bacterial strains and the maintenance of bacterial cultures. An exercise dealing with special media prepares the student for the next part.

Part 4, Microbial Metabolism, includes five exercises on bacterial metabolism that provide the tools for Exercises 8, 32, and 50, Unknown Identifications. These exercises are also useful for learning biochemistry. Exercise 18, Rapid Identification Methods, demonstrates computerization of laboratory work.

Part 5, Growth of Microorganisms, deals with the effects on growth of environmental conditions such as temperature and the presence of oxygen. Current clinical techniques are used to culture anaerobes in Exercise 19. A fast-growing species produces a growth curve in one hour in Exercise 20.

Part 6, Control of Microbial Growth, provides practical applications of concepts of microbial growth. Methods of controlling unwanted microbes in food or in a clinical environment are examined in this part.

Part 7, Microbial Genetics, is of particular interest because of recent notable advances in this field. Exercise 27 demonstrates the isolation of a bacterial mutant. Transformation by an antibiotic-resistance plasmid is performed in Exercise 29. Restriction enzyme digestion and agarose gel electrophoresis are used to analyze the plasmid. Exercise 30 introduces the use of *Agrobacterium* in genetic engineering. In Exercise 31, suspected chemical carcinogens are tested by the Ames Test.

Part 8, The Microbial World, examines the diversity of microorganisms. Information and techniques learned in previous exercises are used to identify a bacterial unknown in Exercise 32. The morphology and ecological niches of free-living eucaryotic organisms studied in microbiology are examined in Exercises 33 through 36. Cyanobacteria are included in Exercise 35, Algae, since they occupy the same habitats and niches as eucaryotic algae.

Part 9, Viruses, provides an opportunity to isolate, cultivate, and quantify bacteriophages. Students determine the host range of a plant virus in Exercise 38.

Part 10, Interaction of Microbe and Host, introduces basic concepts of epidemiology. Methods of tracking and identifying causes of infectious diseases are practiced in the exercises in this part.

Part 11, Immunology, covers the host's response to infectious disease with exercises on nonspecific resistance, serological tests, and fluorescent-antibody testing.

Part 12, Microorganisms and Disease, emphasizes procedures employed in the clinical laboratory. A newly developed technique for determining susceptibility to caries is used in Exercise 47. Exercises on normal and pathogenic bacteria in the human body are followed by identification of an unknown from a simulated clinical sample.

Part 13, Microbiology and the Environment, includes standard methods for the examination of food and water for microbiological quality. Exercise 54, Microbes Used in the Production of Foods, offers an opportunity to study food microbiology. Exercise 55, Microbes in Soil: The Nitrogen Cycle, provides an example of biogeochemical cycles and soil microbiology. Exercise 56, Microbes in Soil: Bioremediation, is a unique exercise illustrating the use of bacteria for cleaning up environmental pollution.

Several appendices at the end of the manual provide a convenient reference to techniques required in several exercises.

Organization of Each Exercise

The exercises in this manual may involve mastering a skill or procedure or understanding a particular concept. Most of the exercises are investigative by design, and the student is asked to analyze the experimental results and draw conclusions. The exercises are organized as follows:

- Each exercise begins with a section called **Objectives,** which lists skills or concepts to be mastered in that exercise. The objectives can be used to test mastery of the new material after completing the exercise.
- The **Background** provides definitions and explanations for each exercise. The student is expected to refer to a textbook for more detailed explanations of the concepts introduced in the laboratory exercises.
- **Materials** lists include supplies and cultures needed for the exercise.
- **Techniques Required** gives any necessary cross-references to earlier exercises.
- **Procedures** are step-by-step instructions, stated as simply as possible and frequently supplemented with diagrams. Questions are occasionally asked in the Procedures section to remind the student of the rationale for a step.
- The **Laboratory Report** provides a space to record results. Questions in the Laboratory Report ask for interpretation of results. In most instances, the results for each student team will be unique; they can be compared with the information given in the Background and other references but will not be identical to these references. The questions are designed to lead the student from a collection of data or observations to a conclusion. The range of questions in each exercise requires students to think about their results, recall facts, and then *use* this information to answer questions.

Illustrations

This book is generously illustrated with diagrams, drawings, and photographs, including 8 pages of photographs in full color. There is a photo quiz in the color section, in which students are asked to identify an unknown. In all cases, the illustrations have been designed as learning aids to be *used* by the student.

Keyware© Identification Software

Keyware 1.0, a software identification program, is available free to adopters of *Laboratory Experiments in Microbiology,* Third Edition. Created by Professor Dennis Lye of Northern Kentucky University, this Hypercard-based program has been designed to be used with a dichotomous key in the identification of unknown isolates. Students can use *Keyware* 1.0 to tentatively identify a possible genus name for their unknowns as part of their regularly scheduled biochemical laboratories (see *Part 4, Microbial Metabolism*). Instructors can modify the program to meet the needs of a particular course. *Keyware* 1.0 will run on any Apple Macintosh computer with one megabyte of memory or more. Users of *Laboratory Experiments in Microbiology* are granted permission to reproduce the minimum number of copies needed for their students.

Instructor's Guide

The comprehensive *Instructor's Guide* (ISBN 0-8053-8493-6) provides the instructor with all the information needed to set up and teach a laboratory course with this manual. It includes:

- General instructions for setting up the lab.
- Information on obtaining and preparing cultures, media, and reagents.
- A master table showing the techniques required for each exercise.
- Cross-references for each exercise to specific pages in *Microbiology: An Introduction,* Fourth Edition.
- For each exercise: helpful suggestions, detailed lists of materials needed, and answers to all the questions in the student manual.

To make *Laboratory Experiments in Microbiology: Brief Edition,* Third Edition easy to use, the experiments have been carefully designed to use inexpensive, readily available, nonhazardous materials. The exercises have been thoroughly tested in our classes in California and Minnesota by students with a wide variety of talents and interests. Our students have enjoyed their microbiology laboratory experiences; we hope yours will, too!

Acknowledgments

We are most grateful to the following individuals for their time, talent, and interest in our work. Each person carefully read and edited critical parts of the manuscript.

Chuck Hoover of the University of California, San Francisco, for making us aware of the new techniques used in dental microbiology for Exercise 47.

Anne Jayne of the University of San Francisco for offering suggestions and a capsule stain.

Patricia Carter of Skyline College for her invaluable assistance in preparing materials and proofreading.

We are indebted to Skyline College and St. Olaf College for providing the facilities and resources in which innovative laboratory exercises can be developed.

We would like to commend the staff at Benjamin/Cummings for their support. In particular, we thank Betsy Dilernia for her attention to detail in copy editing and Larry Olsen for expertly guiding this manual through the production process.

Last, but not least, our gratitude goes to Don Biederman, who provided timely encouragement and unfailing support, and Michelle Johnson, who gave her professional insights and was a sustaining presence.

Reviewers

Robert B. Boley, University of Texas at Arlington
William H. Coleman, University of Hartford
Sidney A. Crow, Jr., Georgia State University
Judith Kandel, California State University at Fullerton
Ralph Reiner, College of the Redwoods

A Special Note to Students

This book is for you. The study of microbiology is dynamic because of the diversity of microbes and the variability inherent in every living organism. Outside of the laboratory — on a forest walk or tasting a fine cheese — we experience the activities of microbes. We want to share our excitement for studying these small organisms. Enjoy!

Ted R. Johnson
Christine L. Case

Contents

Introduction

*Life would not long remain possible in the
absence of microbes.*

LOUIS PASTEUR

Welcome to microbiology! Microorganisms are all around us, and, as Pasteur pointed out a century ago, they play vital roles in the ecology of life on earth. In addition, some microorganisms provide important commercial benefits through their use in the production of chemicals (including antibiotics) and certain foods. Microorganisms are also major tools in basic research in the biological sciences. Finally, as we all know, some microorganisms cause disease—in humans, other animals, and plants.

In this course, you will have first-hand experience with a variety of microorganisms. You will learn the techniques required to identify, study, and work with them. Before getting started, you will find it helpful to read through the suggestions on the next few pages.

Suggestions To Help You Begin

1. Science has a vocabulary of its own. New terms will be introduced in **boldface** throughout this manual. To develop a working vocabulary, make a list of these new terms and their definitions.
2. Because microbes are not visible without a microscope, common names have not been given to them. The word *microbe*, now in common use, was introduced in 1878 by Charles Sedillot. The microbes used in the exercises in this manual are referred to by their *scientific names*. The names will be unfamiliar at first, but do not let that deter you. Practice saying them aloud. Most scientific names are taken from Latin and Greek roots. If you become familiar with these roots, the names will be easier to remember.
3. Microbiology usually provides the first opportunity undergraduate students have to experiment with *living organisms*. Microbes are relatively easy to grow and lend themselves to experimentation. Since there is variability in any population of living organisms, not all the experiments will "work" as

the textbook says. The following exercise will illustrate what we mean:

Write a description of *Homo sapiens* for a visitor from another planet: _____
_____

After you have finished, look around you. Do all your classmates fit the description exactly? Probably not. Moreover, the more detailed you make your description, the less conformity you will observe. During lab, you will make a very detailed description of an organism and probably find that this description does not match your reference exactly.

4. Microorganisms must be cultured or grown to complete most of the exercises in this manual. Accurate record keeping is therefore essential. Cultures will be set up during one laboratory period and will be examined for growth at the next laboratory period. Mark the steps in each exercise with a bright color or a bookmark, so you can return to complete your Laboratory Report on that exercise. *Accurate records* and *good organization* of laboratory work will enhance your enjoyment and facilitate your learning.
5. *Observing* and *recording* your results carefully are the most important parts of each exercise. Ask yourself the following questions for each experiment:
    What did the results indicate?
    Are they what I expected? If not, what happened?
6. If you do not master a technique, try again. In most instances, you will need to utilize the technique again later in the course.
7. Be sure you can answer the questions that are asked in the Procedure for each exercise. These questions are included to reinforce important points that will ensure a successful experiment.
8. Finally, carefully study the general procedures and safety precautions that follow.

General Procedures in Microbiology

In many ways, working in a microbiology laboratory is like working in the kitchen. As some famous chefs have said:

> *Our years of teaching cookery have impressed upon us the fact that all too often a debutant cook will start in enthusiastically on a new dish without ever reading the recipe first. Suddenly an ingredient, or a process, or a time sequence will turn up, and there is astonishment, frustration, and even disaster. We therefore urge you, however much you have cooked, always to read the recipe first, even if the dish is familiar to you. . . . We have not given estimates for the time of preparation, as some people take half an hour to slice three pounds of mushrooms, while others take five minutes.* *

1. Read the laboratory exercises *before* coming to class.
2. *Plan* your work so that all experiments will be completed during the assigned laboratory period. A good laboratory student, like a good cook, is one who can do more than one procedure at a time — that is, one who is efficient.
3. Use only the *required* amounts of materials, so that everyone can do the experiment.
4. *Label* all of your experiments with your name, date, and lab section.
5. Even though you will do most exercises with another student, you must become familiar with all parts of each exercise.
6. Keep *accurate* notes and records of your procedures and results so you can refer to them for future work and tests. Many experiments are set up during one laboratory period and observed for results at the next laboratory period. Your notes are essential to ensure that you perform all the necessary steps and observations.
7. *Demonstrations* will be included in some of the exercises. Study the demonstrations and learn the content.
8. Let your instructor know if you are color-blind; many techniques require discrimination of colors.
9. Keep your cultures current; discard old experiments.
10. *Clean up* your work area when you are finished. Leave the laboratory clean and organized for the next student. Remember:

    Stain and reagent bottles should be returned to their original location.

*J. Child, L. Bertholle, and S. Beck. *Mastering the Art of French Cooking, vol. 1.* New York: Knopf, 1961.

Slides should be washed and put back into the box clean.

All markings on glassware (e.g., Petri plates and test tubes) should be removed before putting glassware into the marked discard trays.

Glass Petri plates should be placed agar-side down.

Swabs and pipettes should be placed in the appropriate disinfectant jars.

Disposable plasticware should be placed in marked autoclave containers.

Used paper towels should be discarded.

Safety in the Laboratory

During your microbiology course, you will learn how to safely handle fluids containing microorganisms. Through practice you will be able to perform experiments so that bacteria and viruses remain in the desired containers and to prevent microbes in the environment from participating in your experiments. These techniques, called **aseptic techniques,** will be a vital part of your work if you are going into health care or into biotechnology.

1. Do not eat, drink, smoke, store food, or apply cosmetics in the laboratory.
2. Wear shoes at all times in the laboratory.
3. Disinfect work surfaces at the beginning and end of every lab period and after every spill. The disinfectant used in this laboratory is _____ .
4. Use mechanical pipetting devices; do not use mouth pipetting.
5. Place a disinfectant-soaked paper towel on the desk while pipetting.
6. Wash your hands after every laboratory exercise. Because bar soaps may become contaminated, use liquid or powdered soaps.
7. Wash your hands immediately and thoroughly if contaminated with microorganisms.
8. Cover spilled microbial cultures with paper towels, and disinfect the towels. Leave for 20 minutes, then clean up the spill.
9. Do not touch broken glassware with your hands; use a broom and dustpan. Place broken glassware contaminated with microbial cultures or body fluids in the To Be Autoclaved container. (See below, Specific Hazards, for broken glassware that is not contaminated.)
10. Place glassware and slides contaminated with blood, urine, and other body fluids in disinfectant.
11. Tie long hair back.
12. Don't perform unauthorized experiments.
13. Don't use equipment without instruction.
14. Horseplay will not be tolerated in the laboratory.

> **Procedures marked with this biohazard icon should be performed carefully, to minimize the risk of transmitting disease.**

15. Work only with your own body fluids and wastes in exercises requiring saliva, urine, blood, or feces, to prevent transmission of disease. The Centers for Disease Control (CDC) state that "epidemiologic evidence has implicated only blood, semen, vaginal secretions, and breast-milk in transmission of HIV" [*MMWR* 36(S2), 8/21/87].

16. If you got this far in the instructions, you'll probably do well in lab. Enjoy lab, and make a new friend.

Specific Hazards

> **Procedures marked with this safety icon should be performed carefully, to minimize risk of exposure to chemicals or fire.**

Alcohol

Keep containers of alcohol away from open flames. When it is necessary to heat alcohol, pour the alcohol into a heat-resistant container and heat on a hot plate.

Glassware Not Contaminated with Microbial Cultures

1. If you break a glass object, sweep up the pieces with a broom and dustpan. Do not pick up pieces of broken glass with your bare hands.
2. Place broken glass in one of the containers marked for this purpose. The one exception to this rule concerns broken thermometers; consult your instructor if you break a thermometer.

Electrical Equipment

1. The basic rule to follow is: Electricity and water don't mix. Do not allow water or any water-based solution to come into contact with electrical cords or electrical conductors. Make sure your hands are dry when you handle electrical connectors.
2. If your electrical equipment crackles, snaps, or begins to give off smoke, do not attempt to disconnect it. Call your instructor immediately.

Fire

1. If *gas burns* from a leak in the burner or tubing, turn off the gas.
2. If you have a *smoldering sleeve,* run water on the fabric.
3. If you have a *very small fire,* the best way to put it out is to smother it with a towel or book (not your hand). Smother the fire quickly.
4. If a *larger fire* occurs, such as in a wastebasket or sink, use one of the fire extinguishers in the lab to put it out. Your instructor will demonstrate the use of the fire extinguishers.
5. In case of a *large fire* involving the lab itself, evacuate the room and building according to the following procedure:
   a. Turn off all gas burners and unplug electrical equipment.
   b. Leave the room and proceed _____.
   c. It is imperative that you assemble in front of the building so that your instructor can take roll to determine if anyone is still inside. Do not wander off.

Accidents and First Aid

1. Report all accidents immediately. Your instructor will administer first aid as required.
2. For spills in or near the eyes, use the eyewash.
3. For large spills on your body, use the safety shower.
4. For heat burns, chill the affected part with ice as soon as possible.

Power Outage

If the electricity goes off, be sure to turn your gas jet off. When the power is restored, the gas will come back on.

Earthquake

Turn off your gas jet and get under your lab desk during an earthquake. Your instructor will give any necessary evacuation instructions.

Orientation Walkabout

Locate the following items in the lab:

Broom and dustpan

Eyewash

Fire blanket

Fire extinguisher

First aid cabinet

Fume hood

Instructor's desk

Reference books

Safety shower

To Be Autoclaved area

Special Practices

1. Keep laboratory doors closed when experiments are in progress.
2. The instructor controls access to the laboratory

and allows access only to people whose presence is required for program or support purposes.

3. Place contaminated materials that are to be decontaminated at a site away from the laboratory into a durable, leakproof container that is closed before being removed from the laboratory.

4. An insect and rodent control program is in effect.

5. A needle should not be bent, replaced in the sheath, or removed from the syringe following use. Place the needle and syringe promptly in a puncture-resistant container and decontaminate, preferably by autoclaving, before discarding or reusing.

6. Inform your instructor if you are pregnant, are taking immunosuppressive drugs, or have any other medical condition (e.g., diabetes, immunologic defect) that might necessitate special precautions in the laboratory.

7. Potential pathogens used in the exercises in this manual are classified in Class 1 by the U.S. Public Health Service. These bacteria present minimal hazard and require ordinary aseptic handling conditions. No special competence or containment is required. These organisms are:

Corynebacterium species
Enterobacter species
Mycobacterium species
Proteus species
Pseudomonas aeruginosa
Salmonella species
Serratia marcescens
Shigella species
Staphylococcus species
Streptococcus species

Laboratory Facilities

1. Interior surfaces of walls, floors, and ceilings are water resistant so that they can be easily cleaned.

2. Bench tops are impervious to water and resistant to acids, alkalis, organic solvents, and moderate heat.

3. Windows in the laboratory are closed and sealed.

4. An autoclave for decontaminating laboratory wastes is available, preferably within the laboratory.

Contact with Blood and Other Body Fluids

The following procedures* should be used by all health-care workers, including students, whose activi-

ties involve contact with patients or with blood or other body fluids. While these procedures were developed by the CDC to minimize the risk of transmitting HIV in a health-care environment, adherence to these guidelines will minimize transmission of *all* nosocomial infections.

1. Wear gloves for touching blood and body fluids, mucous membranes, or non-intact skin and for handling items or surfaces soiled with blood or body fluids. Change gloves after contact with each patient.

2. Wash hands and other skin surfaces immediately and thoroughly if contaminated with blood or other body fluids.

3. Wear masks and protective eyewear or face shields during procedures that are likely to generate droplets of blood or other body fluids.

4. Wear gowns or aprons during procedures that are likely to generate splashes of blood or other body fluids.

5. Wash hands and other skin surfaces immediately after gloves are removed.

6. Mouthpieces, resuscitation bags, or other ventilation devices should be available for use in areas in which the need for resuscitation is predictable. Emergency mouth-to-mouth resuscitation should be minimized.

7. Health-care workers who have exudative lesions or weeping dermatitis should refrain from all direct patient care and from handling patient-care equipment.

8. Pregnant health-care workers are not known to be at greater risk of contracting HIV infection than health-care workers who are not pregnant; however, if a health-care worker develops HIV infection during pregnancy, the infant is at risk of infection. Because of this risk, pregnant health-care workers should be especially familiar with and strictly adhere to precautions to minimize the risk of HIV transmission.

9. In a laboratory exercise where human blood is utilized, students should wear gloves or work only with their own blood and should dispose of all slides and blood-contaminated materials immediately after use. Any cuts or scrapes on the skin should be covered with a sterile bandage.

*Centers for Disease Control. "Recommendations for Prevention of HIV Transmission in Health-Care Settings." *MMWR* 38(S2), 1989.

PART 1

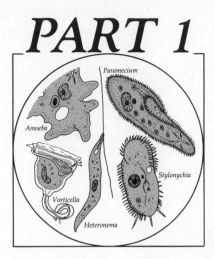

Microscopy

EXERCISES

1 Use and Care of the Microscope

2 Examination of Living Microorganisms

The microscope is a very important tool for a microbiologist. Microscopes and microscopy (microscope technique) are introduced in Exercises 1 and 2, which are designed to help you become familiar with and proficient in the use of the compound light microscope. This knowledge will be valuable in later exercises.

Beginning students frequently become impatient with the microscope and forgo this opportunity to practice and develop their observation skills. Simple observation is a critical part of any science. Making discoveries by observation requires *curiosity* and *patience*. We cannot provide procedures for observation, but we can offer this suggestion: Make *careful sketches* to enhance effective observation. You need not be an artist to draw what you see. In your drawings, pay special attention to the following:

1. Size relationships. For example, how big is a bacterium relative to a protozoan?

2. Spatial relationships. For example, where is one bacterium in relation to the others? Are they all together in chains?

3. Behavior. For example, are individual cells moving or are they all flowing in the liquid medium?

4. Sequence of events. For example, were cells active when you first observed them?

Looking at objects through a microscope is not easy at first, but with a little practice you, like Leeuwenhoek, will make discoveries in the microcosms of raindrops (see the figure). Leeuwenhoek wrote in 1684:

> *Tho my teeth are kept usually very clean, nevertheless when I view them in a Magnifying Glass, I find growing between them a little white matter as thick as wetted flour: In this substance tho I do not perceive any motion, I judged there might probably be living creatures.*
>
> *I therefore took some of this flour and mixed it either with pure rain water wherein were no Animals; or else with some of my Spittle (having no air bubbles to cause a motion in it) and then to my great surprise perceived that the aforesaid matter contained very many small living Animals, which moved themselves very extravagantly. Their motion was strong and nimble, and they darted themselves thro the water or spittle, as a Jack or Pike does thro the water.**

*Quoted in E. B. Fred. "Antoni van Leeuwenhoek." *Journal of Bacteriology* 25:1, 1933.

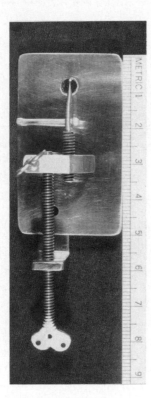

A simple microscope made by Anton van Leeuwenhoek to observe living organisms too small to be seen with the naked eye. The specimen was placed on the tip of the adjustable point and viewed from the other side through the tiny round lens. The highest magnification with his lenses was about 300 ×.

Use and Care of the Microscope

The most important discoveries of the laws, methods and progress of nature have nearly always sprung from the examination of the smallest objects which she contains.

JEAN BAPTISTE LAMARCK

Objectives

After completing this exercise you should be able to:

1. Demonstrate the correct use of a compound light microscope.
2. Diagram the path of light through a compound microscope.
3. Name the major parts of a compound microscope.
4. Identify the three basic morphologies of bacteria.

Background

Virtually all organisms studied in microbiology cannot be seen with the naked eye, but require the use of optical systems for magnification. The microscope was invented shortly before 1600 by either Zacharias Janssen of the Netherlands or Galileo Galilei of Italy. History has been unable to determine which scientist should be given credit. The microscope was not used to examine microorganisms until the 1680s, when a clerk in a dry goods store, Anton van Leeuwenhoek, examined scrapings of his teeth and any other substances he could find. The early microscopes, called **simple microscopes,** consisted of biconvex lenses and were essentially magnifying glasses. To see microbes, a **compound microscope,** which has two lenses between the eye and the object, is required. This optical system magnifies the object, and an illumination system (sun and mirror or lamp) ensures that adequate light is available for viewing. A **brightfield compound microscope,** which has dark objects in a bright field, is used most often.

You will be using a brightfield compound microscope similar to the one in Figure 1.1. The basic frame of the microscope consists of a **base,** a **stage** to hold the slide, an **arm** for carrying the microscope, and a **tube** for transmitting the magnified image. The stage may have two clips or a movable mechanical stage to hold the slide. The light source is in the base. Above the light source is a **condenser,** which consists of several lenses that concentrate light on the slide by focusing it into a cone, as shown in Figure 1.2. The condenser has an **iris diaphragm** that controls the angle and size of the cone of light. This ability to control the *amount* of light ensures that optimal light will reach the image. Above the stage, on one end of the body tube, is a revolving nosepiece holding three or four **objective lenses.** At the upper end of the tube is an **ocular** or **eyepiece lens** ($10 \times$ to $12.5 \times$). If a microscope has only one ocular lens, it is called a **monocular** microscope; a **binocular** microscope has two ocular lenses.

By moving the tube closer to the slide or the stage closer to the objective lens, using the coarse or fine adjustment knobs, one can focus the image. The larger knob, the **coarse adjustment,** is used for focusing with the low-power objectives ($4 \times$ and $10 \times$), and the smaller knob, the **fine adjustment,** is used for focusing high-power and oil immersion lenses. The coarse adjustment knob moves the lenses or the stage longer distances. The area seen through a microscope is called the **field of vision.**

The **magnification** of a microscope depends on the type of objective lens used with the ocular. Compound microscopes have three or four objective lenses mounted on a nosepiece: scanning ($4 \times$), low-power ($10 \times$), high-dry ($40 \times$ to $44 \times$), and oil immersion ($97 \times$ to $100 \times$). The magnification provided by each lens is stamped on the barrel. The total magnification of the object is calculated by multiplying the magnification of the ocular (usually $10 \times$) by the magnification of the objective lens. The most important lens in microbiology is the **oil immersion lens;** it has the highest magnification ($97 \times$ to $100 \times$) and must be used with immersion oil. Optical systems could be built to magnify much more than the $1000 \times$ magnification of your microscope, but the resolution would be poor.

Resolution or **resolving power** refers to the ability of lenses to reveal fine detail or two points distinctly separated. An example of resolution is a car approaching you at night. At first only one light appears, but as it nears, you can distinguish two lights. The resolving power is a function of the wavelength of light used and a characteristic of the lens system called **numerical aperture.** Resolving power is the greatest when two objects are seen as distinct even though they are very close together. Resolving power is expressed in units of length; the smaller the distance, the better the resolving power.

$$\text{Resolving power} = \frac{\text{Wavelength of light used}}{2 \times \text{numerical aperture}}$$

Smaller wavelengths of light increase resolving power. The effect of decreasing the wavelength can be seen by electron microscopes, which utilize electrons as a source of "light." The electrons have an extremely short wavelength and result in excellent resolving power. A light microscope has a resolving power of about 200 nanometers (nm), whereas an electron microscope has a resolving power of less than 0.2 nm. The numerical aperture is engraved on the side of each objective lens (usually abbreviated N.A.). If the numerical aperture increases, for example, from 0.65 to 1.25, the resolving power is improved. The numerical aperture is dependent on the maximum angle of the light entering the objective lens and on the **refractive index** (the amount the light bends) of the material (usually air) between the objective lens and the slide. This relationship is defined by the following:

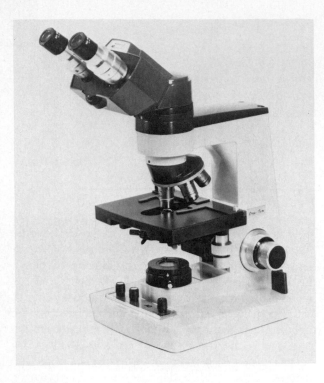

Figure 1.1 _____

Photograph of a compound light microscope.

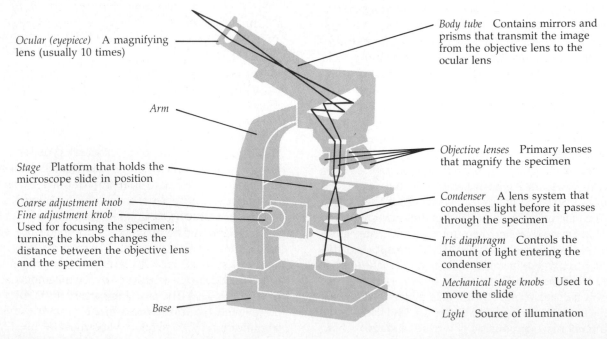

Ocular (eyepiece)   A magnifying lens (usually 10 times)

Arm

Stage   Platform that holds the microscope slide in position

Coarse adjustment knob
Fine adjustment knob
Used for focusing the specimen; turning the knobs changes the distance between the objective lens and the specimen

Base

Body tube   Contains mirrors and prisms that transmit the image from the objective lens to the ocular lens

Objective lenses   Primary lenses that magnify the specimen

Condenser   A lens system that condenses light before it passes through the specimen

Iris diaphragm   Controls the amount of light entering the condenser

Mechanical stage knobs   Used to move the slide

Light   Source of illumination

Figure 1.2 _____

The compound light microscope: its principal parts and their functions. Lines from the light source through the ocular lens illustrate the path of light.

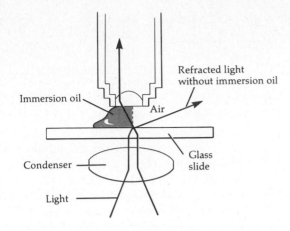

Figure 1.3 _____

Refractive index. Because the refractive indexes of the glass microscope slide and immersion oil are the same, the oil keeps the light rays from refracting.

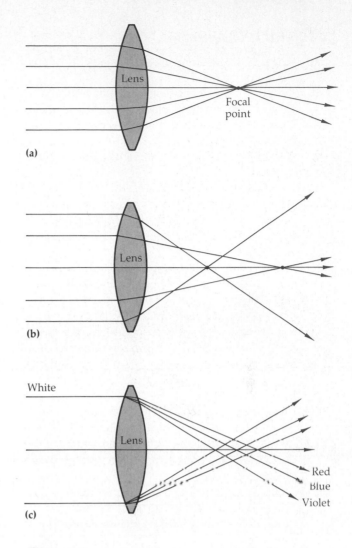

(a)

(b)

(c)

Figure 1.4 _____

Focal point. **(a)** An image is formed when light converges at one point, called the focal point. **(b)** Spherical aberration. Curved lenses result in light passing through one region of the lens having a different focal point than light passing through another part of the lens. **(c)** Chromatic aberration. Each wavelength of light may be given a different focal point by the lens.

N.A. = $N \sin \theta$
   N = Refractive index of medium
   θ = Angle between the most divergent light ray
        gathered by the lens and the center of the lens

As shown in Figure 1.3, light is refracted when it emerges from the slide because of the change in media as the light passes from glass to air. When immersion oil is placed between the slide and the oil immersion lens, the light ray continues without refraction because immersion oil has the same refractive index (N = 1.52) as glass (N = 1.52). This can be seen easily. When you look through a bottle of immersion oil, you cannot see the glass rod in it because of the identical N values of the glass and immersion oil. The result of using oil is that light loss is minimized, and the lens focuses very close to the slide.

As light rays pass through a lens they are bent to converge at the **focal point,** where an image is formed (Figure 1.4*a*). When you bring the center of a microscope field into focus, the periphery may be fuzzy due to the curvature of the lens, resulting in multiple focal points. This is called **spherical aberration** (Figure 1.4*b*). Spherical aberrations can be minimized by the use of the iris diaphragm, which eliminates light rays to the periphery of the lens, or by a series of lenses resulting in essentially a flat optical system. Sometimes a multitude of colors, or **chromatic aberration,** is seen in the field (Figure 1.3*c*). This is due to the prismlike effect of the lens as various wavelengths of white light pass through to a different focal point for each wavelength. Chromatic aberrations can be minimized by the use of filters (usually blue); or by lens systems corrected for red and blue light, called achromatic lenses; or by lenses corrected for red, blue, and other wavelengths,

called apochromatic lenses. The most logical, but most expensive, method of eliminating chromatic aberrations is to use a light source of one wavelength, or **monochromatic light.**

Compound microscopes require a light source. The light may be reflected to the condenser by a mirror under the stage. If your microscope has a mirror, the sun or a lamp may be used as the light source. Most newer compound microscopes have a built-in illuminator in the base. The *intensity* of the light can often be adjusted with a transformer or rheostat.

The microscope is a very important tool in microbiology, and it must be used carefully and correctly. Follow these guidelines *every* time you use a microscope.

General Guidelines

1. Carry the microscope with one hand beneath the base and one hand on the arm.
2. Do not tilt the microscope; instead, adjust your stool so you can comfortably use the instrument.
3. Observe the slide with both eyes open, to avoid eye strain.
4. Always focus by moving the lens away from the slide.
5. Always focus slowly and carefully.
6. When using the low-power lens, the iris diaphragm should be barely open so that good contrast is achieved. More light is needed with higher magnification.
7. Before using the oil immersion lens, have your slide in focus under high power. *Always focus with low power first.*
8. Keep the stage clean and free of oil. Keep all lenses except the oil immersion lens free of oil.
9. Keep all lenses clean. Use *only* lens paper to clean them. Wipe oil off before putting your microscope away. Do not touch the lenses with your hands.
10. Clean the ocular lens carefully with lens paper. If dust is present, it will rotate as you turn the lens.
11. After use, remove the slide, wipe oil off, put the dust cover on, and return the microscope to the designated area.
12. When a problem does arise with the microscope, obtain help from the instructor. Do not use another microscope unless yours is declared "out of action."

Materials

Microscope

Immersion oil

Lens paper

Prepared slides of algae, fungi, protozoans, and bacteria

Techniques Required

Part 1 Introduction.

Procedure

1. Place the microscope on the bench squarely in front of you.
2. Obtain a slide of algae or fungi and place it on the stage.

3. Adjust the eyepieces on a binocular microscope to your own personal measurements.
   a. Look through the eyepieces and, using the thumb wheel, adjust the distance between the eyepieces until one circle of light appears.
   b. With the low-power (10×) objective in place, cover the left eyepiece with a small card and focus the microscope on the slide. When the right eyepiece has been focused, remove your hand from the focusing knobs and cover the right eyepiece. Looking through the microscope with your left eye, focus the left eyepiece by turning the eyepiece adjustment. Make a note of the number at which you focused the left eyepiece so you can adjust any binocular microscope for your own eyes.
4. Raise the condenser up to the stage. On some microscopes, the condenser can be focused by the following procedure:
   a. Focus with the 4× or 10× objective.
   b. Close the iris diaphragm so only a minimum of light enters the objective lens.
   c. Lower the condenser until the light is seen as a circle in the center of the field. On some microscopes the circle of light may be centered (Figure 1.5) using the centering screws found on the condenser.

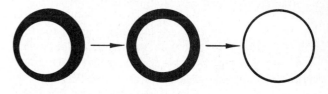

Figure 1.5 _____

Using low power, lower the condenser until a distinct circle of light is visible. Center the circle of light using the centering screws. Open the iris diaphragm until the light just fills the field.

   d. Raise the condenser up to the slide, lower it, and stop when the color on the periphery changes from pink to blue (1 or 2 mm below the stage).
   e. Open the iris diaphragm until the light just fills the field.
5. Diagram some of the cells on the slide under low power. Use a minimum of light by adjusting the _____ .
6. When an image has been brought into focus with low power, rotate the turret to the next lens, and the subject will remain almost in focus. All of the objectives (with the possible exception of the 4×)

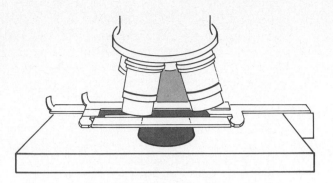

(a) Move the high-dry lens out of position.

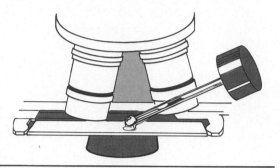

(b) Place a drop of immersion oil in the center of the slide.

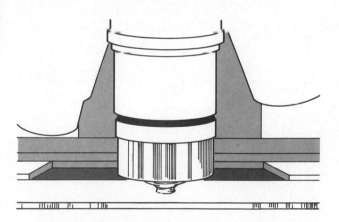

(c) Move the oil immersion lens into position.

Figure 1.6 _____
Using the oil immersion lens.

are **parfocal;** that is, when a subject is in focus with one lens, it will be in focus with all of the lenses. When you have completed your observations under low power, swing the high-dry objective into position and focus. Use the fine adjustment. Only a slight adjustment should be required. Why? _____
_____

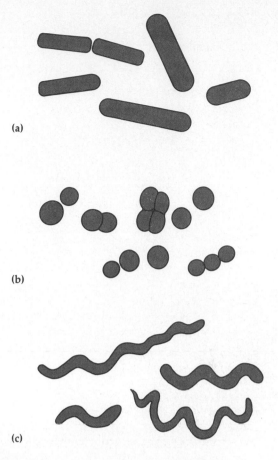

Figure 1.7 _____
Basic shapes of bacteria. **(a)** Bacillus (plural, bacilli), or rod. **(b)** Coccus (plural, cocci). **(c)** Spirillum (plural, spirilla).

More light is usually needed. Again, draw the general size and shape of some cells.

7. Move the high-dry lens out of position and place a drop of immersion oil on the area of the slide you are observing. Carefully click the oil immersion lens into position. It should now be immersed in the oil (Figure 1.6). Fine adjustment should bring the object into focus. Note the shape and size of the cells. Did the color of the cells change with the different lenses? _____
Did the size of the field change? _____

8. When your observations are completed, move the turret to bring a low-power objective into position. *Do not* rotate the high-dry (40×) objective through the immersion oil. Clean the oil off the objective lens with lens paper, and clean off the slide with tissue paper or a paper towel. Remove the slide. Repeat this procedure with all the available slides. When observing the bacteria, note the three different morphologies, or shapes, shown in Figure 1.7.

Examination of Living Microorganisms

Objectives

After completing this exercise you should be able to:

1. Prepare and observe wet mount slides and hanging-drop slides.
2. Recognize different types of microbes in unstained preparations.
3. Distinguish between true motility and Brownian movement.

Background

Anton van Leeuwenhoek was the first known individual to observe living microbes in a suspension. Unfortunately, he was very protective of his homemade microscopes and left no descriptions of how to make them. During his lifetime he kept "for himself alone" his microscopes and his method of observing "animalcules." Leeuwenhoek made a new microscope for each specimen. Directions for making a replica of Leeuwenhoek's microscope can be found in *American Biology Teacher*.*

In this exercise you will examine, using wet mount techniques, different fluid environments to help you become aware of the numbers and varieties of microbes found in nature. The microbes will exhibit either Brownian movement or true motility. **Brownian movement** is not true motility but rather is movement caused by the molecules in the liquid striking an object and causing the object to shake or bounce. In Brownian movement the particles and microorganisms all vibrate at about the same rate and maintain their relative positions. Motile microorganisms move from one position to another. Their movement appears more directed than Brownian movement, and occasionally the cells may roll or spin.

Many kinds of microbes, such as protozoans, algae, and bacteria, can be found in pond water and in infusions of organic matter. Direct examination of living microorganisms is very useful in determining size, shape, and movement. A wet mount is a fast way to observe bacteria. Motility and larger microbes are more easily observed in the greater depth provided by a hanging drop. Evaporation of the suspended drop of fluid is reduced by a petroleum jelly seal.

*W. G. Walter and H. Via. "Making a Leeuwenhoek Microscope Replica." *American Biology Teacher* 30 (6):537–539, 1968.

Materials

Slides

Cover slips

Hanging-drop (depression) slide

Petroleum jelly

Pasteur pipettes

Cultures

Hay infusion, incubated 1 week in light

Hay infusion, incubated 1 week in dark

Peppercorn infusion

12- to 18-hour-old broth culture of *Bacillus*

Techniques Required

Exercise 1.

Procedure

Wet Mount Technique

1. Suspend the infusions by stirring or shaking carefully. Transfer a small drop of one hay infusion, using a Pasteur pipette, to a slide.
2. Handle the cover slip carefully by its edges and place it on the drop.
3. Gently press on the cover slip with the end of a pencil.
4. Place the slide on the microscope stage and observe with low power. Adjust the iris diaphragm so a small amount of light is admitted. Concentrate your observations on the larger, more rapidly moving organisms. At this magnification, bacteria are barely discernible as tiny dots. Figure 2.1 may be helpful in identifying some of the microorganisms.
5. Examine with the high-dry lens, then increase the light and focus carefully. Bacteria should now be magnified sufficiently to see them.
6. After recording your observations, examine the slide under oil immersion. Some microorganisms are motile, while others exhibit Brownian movement.

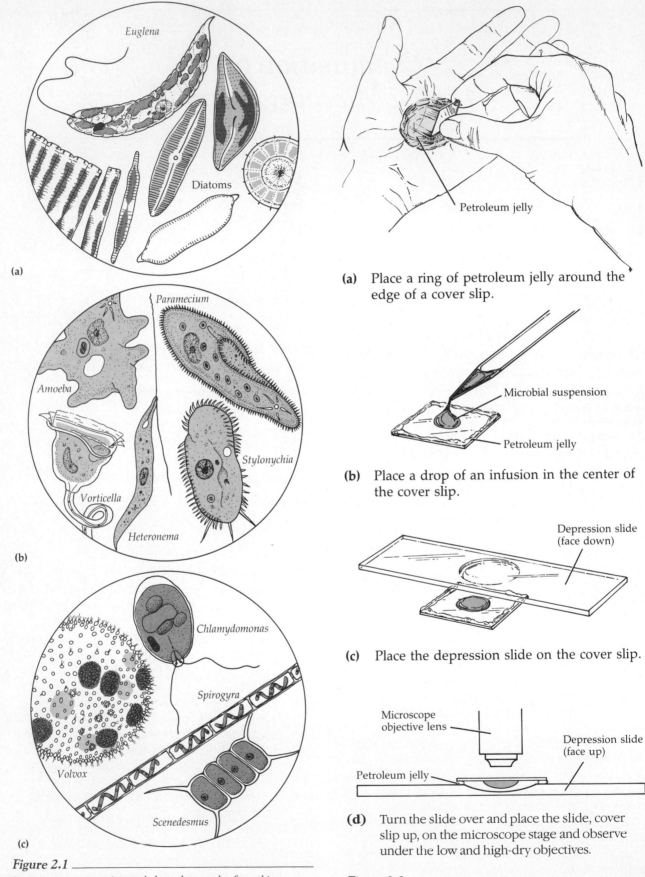

(a) Place a ring of petroleum jelly around the edge of a cover slip.

(b) Place a drop of an infusion in the center of the cover slip.

(c) Place the depression slide on the cover slip.

(d) Turn the slide over and place the slide, cover slip up, on the microscope stage and observe under the low and high-dry objectives.

Figure 2.1 _____

Some common protists and algae that can be found in infusions. **(a)** Algal-like protists. **(b)** Animal-like protists. **(c)** Algae.

Figure 2.2 _____

Hanging-drop preparation.

7. If you want to observe the motile organisms further, place a drop of alcohol or Gram's iodine at the edge of the cover slip and allow it to run under and mix with the infusion. What does the alcohol or iodine do to these organisms? _____ They can now be observed more carefully.

8. Record your observations, noting the relative size and shape of the organisms.

9. Make a wet mount from the other hay infusion, and observe, using the low and high-dry objectives. Record your observations.

10. Clean all the slides and return them to the slide box. Cover slips can be discarded in the disinfectant jar.

Hanging-Drop Procedure

1. Obtain a depression slide.

2. Place a small amount of petroleum jelly on the palm of your hand and smear it into a circle about the size of a 50-cent coin.

3. Pick up a cover slip (by its edges) and carefully scrape the petroleum jelly with an edge of the cover slip to get a small rim of petroleum jelly. Repeat with the other three edges (Figure 2.2*a*), keeping the petroleum jelly on the same side of the cover slip.

4. Place the cover slip on a paper towel, with the petroleum jelly side up.

5. Transfer a drop of the peppercorn infusion to the cover slip.

6. Place a slide over the drop and quickly invert so the drop is suspended (Figure 2.2*b* and *c*). Why should the drop be hanging? _____
_____

7. Examine under low power (Figure 2.2*d*) by locating the edge of the drop and moving the slide so the edge of the drop crosses the center of the field.

8. Reduce the light with the iris diaphragm and focus. Observe the different sizes, shapes, and types of movement.

9. Switch to high-dry and record your observations. Do not focus down. Why not? _____
_____

10. When finished, clean your slide, and, using a new cover slip, repeat the procedure with the culture of *Bacillus*. Record your observations.

11. Wipe the oil from the objective lens with lens paper and return your microscope to its proper location. Clean your slides well and return them.

I.1 Gram-negative rods.
Escherichia coli (Exercise 5).

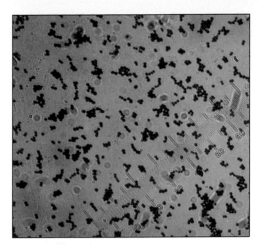

I.2 Gram-positive cocci. *Staphylococcus aureus* (Exercise 5).
One ocular micrometer division equals 1 μm.

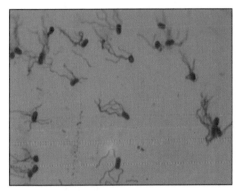

I.3 Flagella stain. Peritrichous flagella of
Proteus vulgaris (Exercise 7).

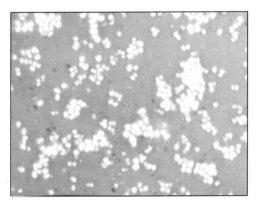

I.4 Negative stain.
Staphylococcus aureus (Exercise 4).

II.1 Crown gall caused by
Agrobacterium tumefaciens on an
ornamental Japanese spindle-tree
(Exercise 30).

**II.2 DNA is collected on a
glass rod** (Exercise 28).

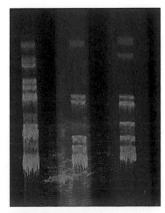

**II.3 Restriction enzyme
digests of a small piece of DNA.**
A different enzyme was used
in each lane (Exercise 29).

PLATE III

III.1 **Reactions in O-F glucose medium.** Bacterial growth is visible along the stab lines in each tube. There is no change in **1**. The blue color in **2** indicates metabolism of peptones. Glucose is fermented (anaerobically) in **3** and oxidized (aerobically) in **4** (Exercise 13).

III.2 **Reactions in fermentation tubes.** **1** is an uninoculated control. Growth and acid production from carbohydrate fermentation are seen in **2**. Acid and gas are produced from fermentation in **3** (Exercise 14).

III.3 **Reactions in litmus milk.** **1** is an uninoculated control. **2** shows an alkaline reaction. Acid is produced in **3**. The litmus is reduced in **4**. The clear fluid layer in **5** is the result of peptonization (Exercise 15).

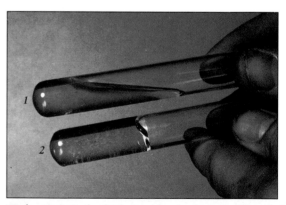

III.4 **Gelatin hydrolysis.** After hydrolysis (**1**), gelatin remains liquid. **2** is unhydrolyzed gelatin. (Exercise 15).

III.5 **Fifteen biochemical tests are performed in Enterotube II.** The bottom tube is uninoculated (Exercise 18).

MICROBIAL METABOLISM

III.6 Methyl red test. Red color (in **1**) after addition of methyl red indicates a positive test. **2** is methyl red-negative (Exercise 14).

III.7 Voges-Proskauer test. A positive Voges-Proskauer test develops a wine-red color when exposed to oxygen (Exercise 14).

III.8 Citrate test. Utilization of citric acid as the sole carbon source in Simmons citrate agar causes the indicator to turn blue (**1**). **2** is citrate-negative (Exercise 18).

1   2

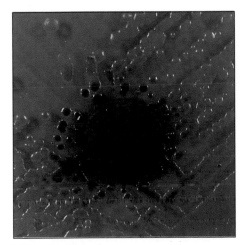

III.9 Oxidase test. Colonies of cytochrome oxidase-positive bacteria turn black as oxidase reagent diffuses from the oxidase test disk into the nutrient agar. (Exercise 17).

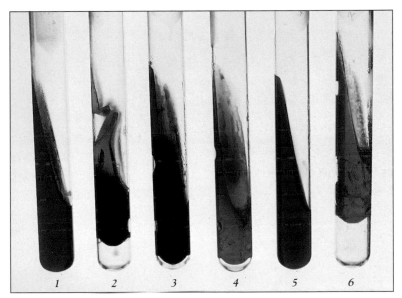

1   2   3   4   5   6

III.10 Reactions in triple sugar iron (TSI) agar. 1 shows growth with no fermentation or hydrogen sulfide (H_2S) production. **2** shows blackening due to H_2S and acid and gas from fermentation of glucose and sucrose and/or lactose. Sucrose and lactose were not fermented in **3**; H_2S production masks the glucose fermentation reaction although gas is produced. Acid and gas are produced from glucose in **4**. **5** is uninoculated. **6** shows acid and gas production from glucose and sucrose and/or lactose (Exercise 48).

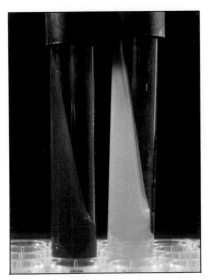

III.11 Urease production. Hydrolysis of urea produces ammonia, which turns the indicator fuchsia (Exercise 15).

III.12 Ornithine decarboxylation. Removal of CO_2 from ornithine turns the indicator purple (Exercise 16).

III.13 Phenylalanine deaminase. Removal of the amino group produces an organic acid that forms a green complex with the ferric ion-containing indicator (Exercise 16).

PLATE IV   EUKARYOTES

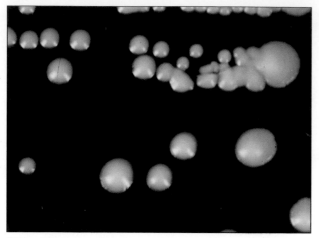

IV.1   The yeast *Saccharomyces cerevisiae* produces circular, convex, glistening colonies on Sabouraud dextrose agar (7000×) (Exercise 33).

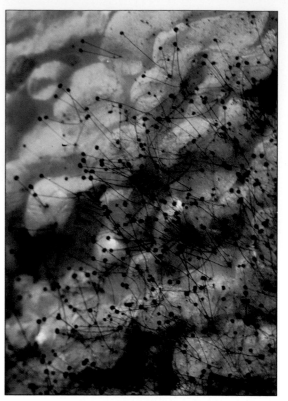

IV.2   *Rhizopus nigricans* growing on styrofoam impregnated with a nutrient medium. Dark sporangia are visible at the tips of the sporangiophores (Exercise 34).

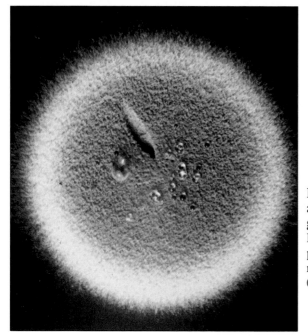

IV.3   *Penicillium* has a white mycelium and green conidiospores. Note the crystallized penicillin on top of the mold colony. (Exercise 34).

IV.4   Red snow at Tioga Pass (Yosemite National Park) due to the green alga *Chlamydomonas nivalis*. The green color of chlorophyll is masked by red carotenoid pigments produced in the presence of intense light (Exercise 35).

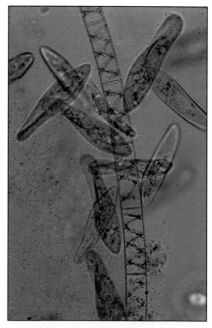

IV.5   The green alga *Spirogyra* and protozoan *Paramecium*, in pond water. (100,000×) (Exercises 35 and 36).

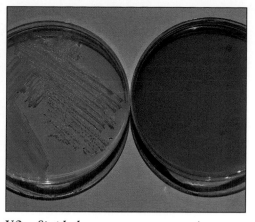

V.2   *Staphylococcus aureus* growing on **mannitol salt agar.** The yellow color indicates that mannitol is fermented. The red plate is uninoculated (Exercise 45).

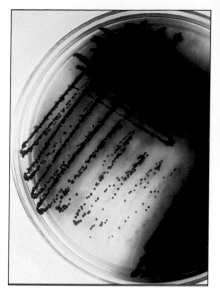

V.1   *Serratia marcescens* colonies on nutrient agar after incubation at 25ºC. Note color variations due to mutations in prodigiosin production (Exercise 27).

V.3   *Pseudomonas aeruginosa* produces a water-soluble blue pigment on Pseudomonas Agar P (Exercise 49).

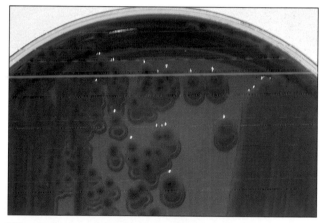

V.4   Colonies of *Escherichia coli* (shown) and *Citrobacter* develop a metallic green sheen on EMB agar (Exercises 48, 49, 51, and 52).

V.5   Colonies of *Enterobacter* have a distinctive blue "fish-eye" appearance on EMB agar (Exercises 48, 49, 51, and 52).

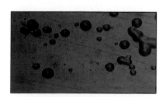

V.6   Nonlactose-fermenters such as *Proteus vulgaris* (shown) produce colorless colonies on EMB agar (30×) (Exercises 48, 49, 51, and 52).

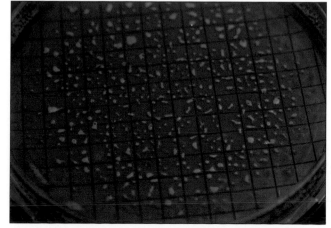

V.7   Coliforms trapped on a membrane filter grow into colonies when the filter is laid on Endo agar and incubated (Exercise 52).

PLATE VI ENVIRONMENTAL MICROBIOLOGY

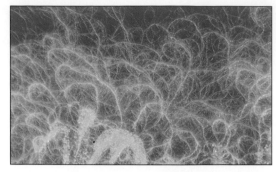

VI.1 *Bacillus cereus* var. *mycoides* colonies. Easily recognized by their filamentous form. Bending of the filaments to the left or right is a strain characteristic (Exercise 11).

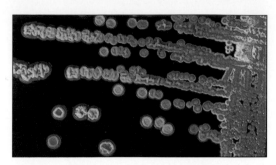

VI.2 Actinomycete colonies. *Streptomyces griseus* colonies penetrate into the agar as well as extend above. Powdery appearance in the corner is due to conidiospores (Exercise 11).

VI.3 Root nodules. *Rhizobium leguminosarum* infected the roots of this vetch plant and caused the production of pink nodules (Exercise 55).

PLATE VII PATHOLOGY

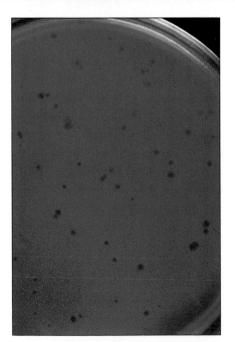

VII.1 Numerous plaques (clearings) in this *Escherichia coli* culture are due to growth of a T-even bacteriophage (Exercise 37).

VII.2 Chlorosis (loss of green color) and plaques (spotting) in this tomato leaf are due to a tobacco mosaic virus infection (Exercise 38).

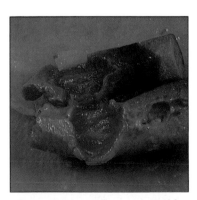

VII.3 Carrot soft rot caused by *Erwinia carotovora* (Exercise 40).

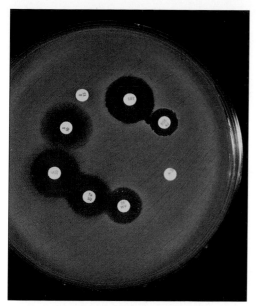

VIII.1   Antibiotic sensitivity of this *Staphylococcus aureus* culture is demonstrated by the Kirby-Bauer agar diffusion test (Exercise 25).

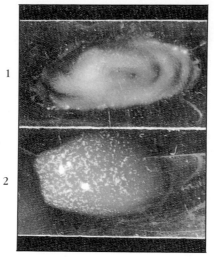

VIII.2   **Coagulase test.** Many pathogenic strains of *Staphylococcus aureus* can coagulate plasma. Bacteria are caught in fibrin clumps in **2** (Exercise 45).

VIII.3   Growth of *Streptococcus pneumoniae* on blood agar demonstrates alpha-hemolysis. Note the greenish color around the colonies (Exercise 46).

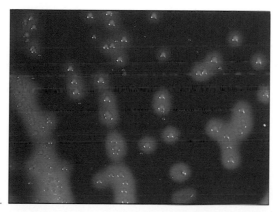

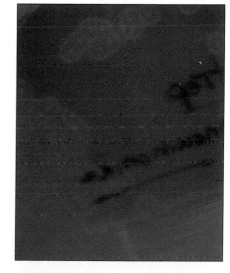

VIII.4   Beta-hemolysis produces a clear area around the colonies of *Streptococcus pyogenes* on blood agar (Exercise 46).

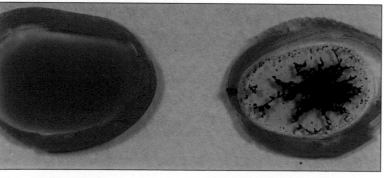

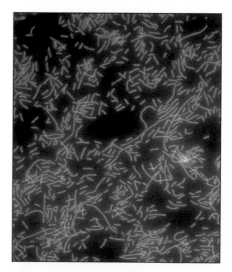

IX.1   **Slide agglutination test for the ABO blood group.** A negative reaction on the left and a positive reaction (indicated by agglutination) on the right (Exercise 42).

IX.2   **Fluorescent-antibody preparation of *Escherichia coli*** (Exercise 44). (Courtesy of Genentech).

PLATE X UNKNOWN IDENTIFICATION-A PHOTO QUIZ

This is an example of how to identify an unknown bacterium.

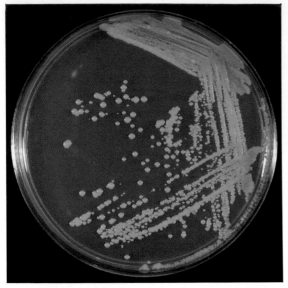

X.1 The unknown organism is isolated in pure culture (Exercise 11).

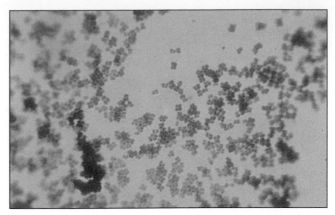

X.2 A Gram stain is made from the pure culture (Exercise 5).

X.3 Based on the result of the Gram stain, a catalase test is indicated (Exercise 17).

X.4 Results of the previous tests lead to the next series of tests. A glucose fermentation tube is inoculated with the unknown bacterium (Exercise 13).

X.5 Lipid hydrolysis test. The light color around the colonies indicates hydrolysis of the lipid in the medium (Exercise 56).

Using the identification keys in Part 12 of this manual, identify the genus and species of this bacterium.

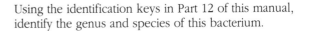
Answer: *Micrococcus luteus*

PART 2

Staining Methods

In 1877, Robert Koch wrote:

> *How many incomplete and false observations might have remained unpublished instead of swelling the bacterial literature into a turbid stream, if investigators had checked their preparations with each other?**

To solve this problem, he introduced into microbiology the procedures of air drying, chemical fixation, and staining with aniline dyes.

In addition to making a lasting preparation, staining bacteria enhances the contrast between bacteria and the surrounding material and permits observation of greater detail and resolution than wet mount procedures (Exercise 2). Microorganisms are prepared for staining by smearing them onto a microscope slide (Exercise 3).

In the exercises in Part 2, bacteria will be transferred from growth or culture media to microscope slides using an inoculating loop. An **inoculating loop** is a nichrome wire held with an insulated handle. Before and after it is used, the inoculating loop is sterilized by heating or **flaming;** that is, the loop is held in the

*Quoted in H. A. Lechevalier and M. Solotorovsky. *Three Centuries of Microbiology.* New York: Dover Publications, 1974, p. 79.

flame of a burner (part *a* of the figure) or electric incinerator (part *b* of the figure) until it is red hot.

Staining techniques may involve **simple stains,** in which only one reagent is used and all bacteria are usually stained similarly (Exercises 3 and 4), or **differential stains,** in which multiple reagents are used and bacteria react to the reagents differently (Exercises 5 and 6). Structural stains are used to identify specific parts of microorganisms (Exercise 7).

Most of the bacteria that are grown in laboratories are rods or cocci (see Figure 1.7). Rods and cocci will be used throughout this part. In Exercise 8, "Morphologic Unknown," you will be asked to determine whether an unknown culture is a rod or a coccus.

(b) The loop can be sterilized in an electric incinerator.

Sterilizing an inoculating loop.

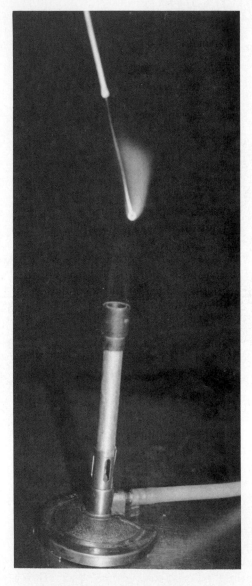

(a) Before and after use, sterilize the loop until it is red hot with a flame.

Preparation of Smears and Simple Staining

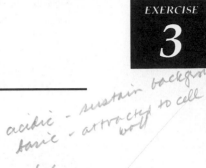

Objectives

After completing this exercise you should be able to:

1. Make and heat fix a smear.
2. List the advantages of staining microorganisms.
3. Explain the basic mechanism of staining.
4. Perform a simple direct stain.

Background

Most stains used in microbiology are synthetic **aniline** (coal tar derivative) dyes derived from benzene. The dyes are usually salts, although a few are acids or bases, composed of charged colored ions. The ion that is colored is referred to as a **chromophore.** For example,

$$\text{Methylene blue chloride} \rightleftharpoons \underset{\text{(Chromophore)}}{\text{Methylene blue}^+} + \text{Cl}^-$$

If the chromophore is a positive ion like methylene blue, the stain is considered a **basic stain;** if a negative ion, it is an **acidic stain.** Most bacteria are stained when a basic stain permeates the cell wall and adheres by weak ionic bonds to the negative charges of the bacterial cell.

Staining procedures that use only one stain are called **simple stains.** A simple stain that stains the bacteria is a **direct stain,** and a simple stain that stains the background, but leaves the bacteria unstained, is a **negative stain** (Exercise 4). Simple stains can be used to determine cell morphology, size, and arrangement.

Before bacteria can be stained, a smear must be made and heat fixed. A **smear** is made by spreading a bacterial suspension on a clean slide and allowing it to air dry. The dry smear is passed through a Bunsen burner flame several times to **heat fix** the bacteria. Heat fixing denatures bacterial enzymes, preventing them from digesting cell parts, which causes the cell to break, a process called autolysis. The heat also enhances the adherence of bacterial cells to the microscope slide.

Materials

Methylene blue ~ *make organism visible*

Wash bottle of distilled water

Slides (2) *clear 1st*

Inoculating loop

acidic - sustain background
basic - attracted to cell wall

colors
basic dyes
heat wire
30 sec.

Cultures

Staphylococcus epidermidis slant

Bacillus megaterium broth

Techniques Required

Exercise 1 and Part 2 Introduction.

Procedure

1. Clean your slides well with abrasive soap or cleanser; rinse and dry. Handle clean slides by the end or edge. Use a marker to make a dime-sized circle on each slide, on the bottom of the slide so they will not wash off. Label each slide according to the bacterial culture used.

2. For the bacterial culture on solid media, place 1 or 2 loopfuls of distilled water in the center of the circle on one slide, using the inoculating loop. Which bacterium is on a solid medium? _____

3. Sterilize your inoculating loop by holding it in the hottest part of the flame (at the edge of the inner blue area) until it is red hot (see the figure on p. 22). The entire wire should get red. Allow the loop to cool so that bacteria picked up with the loop won't be killed. Allow the loop to cool without touching it. Cooling takes about 30 seconds. You will determine the appropriate time with a little practice.

> ☣ **The loop must be cool before inserting it into a medium. A loop that is too hot will spatter the medium and move bacteria into the air.**

4. Smear preparation (Figure 3.1).
   a. Using the cooled loop, scrape a *small* amount of the culture off the slant. If you hear the sizzle of boiling water when you touch the agar with

(a) Mark the smear area with a marking pencil on the underside of a clean slide.

FROM SOLID MEDIUM   FROM LIQUID MEDIUM

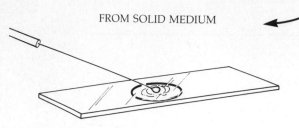

(b) Place 1 or 2 loopfuls of water on the slide.

(d) Place 2 or 3 loopfuls of the liquid culture on the slide with a sterile loop.

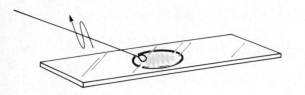

(c) Transfer a very small amount of the culture with a sterile loop. Mix with the water on the slide.

(e) Spread the bacteria within the ring.

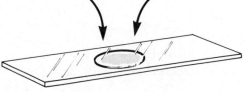

(f) Allow the smear to air dry at room temperature.

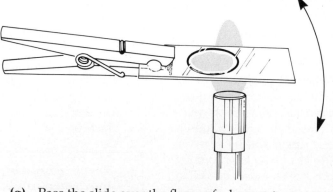

(g) Pass the slide over the flame of a burner two or three times.

Figure 3.1

Preparing a bacterial smear.

the loop, reflame your loop and begin again. Why? _____

Try not to gouge the agar. Emulsify (to a milky suspension) the organism in the drop of water, and spread the suspension to fill a majority of the circle. The smear should look like diluted skim milk. Flame your loop again.

b. Make a smear of bacteria from the broth culture on the other slide. *Do not use water,* because the bacteria are already suspended in water. Flick the tube of broth culture lightly with your finger to resuspend sedimented bacteria, and place 2 or 3 loopfuls of the culture in the circle. Flame your loop between each loopful. Spread the culture within the circle (Figure 3.2).

c. Flame your loop.

> ☣ **Always flame your loop after using it and before setting it down.**

---

d. Let the smears dry. *Do not* blow on the slides, as this will move the bacterial suspension. *Do not* flame the slide, as flaming will distort the cells' shapes.

e. Hold the slides with a clothespin and heat fix the smears by passing the slides quickly through the blue flame two or three times. Do not heat fix until the smear is completely dry. Why? _____
_____

5. Staining.

  a. Do one slide at a time. Use a clothespin to hold the slide, or place it on a staining rack.

  b. Cover the smear with methylene blue and leave for 30 to 60 seconds.

  c. Carefully wash the excess stain off with dis-

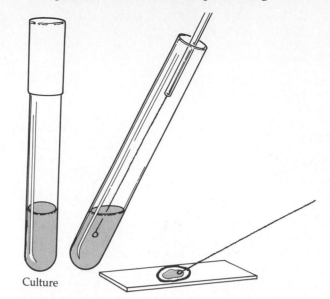

Figure 3.2 _____
A loopful of microbial suspension is transferred to a slide.

Culture

tilled water from a wash bottle. Let the water run down the tilted slide.

  d. Gently blot the smear with a paper towel or absorbent paper and let it dry. *quits running blue*

6. Examine your stained smears microscopically using the low, high-dry, and oil immersion objectives. Put the oil *directly* on the smear; cover slips are not needed. Record your observations with labeled drawings.

7. Blot the oil from the objective lens with lens paper, and return your microscope to its proper location. Clean your slides well, or save them as described in step 8.

8. Stained bacterial slides can be stored in a slide box. Remove the oil from the slide by blotting with a paper towel. Any residual oil won't matter.

EXERCISE 3

Preparation of Smears and Simple Staining

Name _ANNABELLE G. BUENDIA_

Date _JUNE 7, 1993_

Lab Section _____

Purpose _Make and heat fix a smear. List the advantages of staining microorganisms. Explain the basic mechanism of staining. Perform a simple direct stain._

Data

Appearance	*Staphylococcus epidermidis*	*Bacillus megaterium*
Sketch a few bacteria under 1000 × magnification		
Morphology (shape)	coccus	rod
Arrangement of cells relative to one another	close to one another	long chains

Questions

1. Of what value is a simple stain? _Simple stains highlight the entire micro-organism so that cellular shapes and structures are visible._

2. Can dyes other than methylene blue be used for direct staining? _yes_

3. What would happen if no heat fixing were done? _bacteria will not fix or adhere to the slides_

   Or too much heat applied? _bacteria will be destroyed and can't be examined_

EXERCISE 4

Negative Staining

Objectives

After completing this exercise you should be able to:

1. Explain the application and mechanism of the negative stain technique.
2. Prepare a negative stain.

Background

The **negative stain** technique does not stain the bacteria but stains the background. The bacteria will appear clear against a stained background. The stain does not stain the bacteria because of ionic repulsion: The bacteria and the acidic stain both have negative charges.

No heat fixing or strong chemicals are used, so the bacteria are less distorted than in other staining procedures. The negative stain technique is very useful in situations where other staining techniques don't clearly indicate cell morphology or size.

Materials

Nigrosine

Clean slides (6)

Distilled water

Sterile toothpicks

Cultures

Bacillus subtilis

Staphylococcus epidermidis

Techniques Required

Exercises 1 and 3.

Procedure (Figure 4.1)

1. Slides must be clean and grease-free. See Exercise 3, Procedure step 1.
2. **a.** Place a *small* drop of nigrosine at the end of the slide. For cultures on solid media, add a loopful of distilled water and emulsify a small amount of the culture in the nigrosine–water drop. For broth cultures, mix a loopful of the culture into the drop of nigrosine (Figure 4.1*a*). Do not spread the drop or let dry.
   b. Using the end edge of another slide, spread the drop out (Figure 4.1*b* and *c*) to produce a smear varying from opaque black to gray. The angle of the spreading slide will determine the thickness of the smear.
   c. Let the smear air dry (Figure 4.1*d*). *Do not heat fix.*
   d. Prepare a negative stain of the other culture.
3. Examine the stained slides microscopically using the low, high-dry, and oil immersion objectives (Figure 4.2). As a general rule, if a few short rods are seen with small cocci, the morphology is rod-shaped. The apparent cocci are short rods viewed from the end or products of the cell division of small rods. (See color plate I.4.)
4. **a.** Place a small drop of nigrosine and a loopful of water at the end of a slide.
   b. Scrape the base of your teeth and gums with a sterile toothpick.
   c. Mix the nigrosine–water with the toothpick to get an emulsion of bacteria from your mouth.

> ☣ **Discard the toothpick in disinfectant.**

   d. Follow steps 2b and 2c above to complete the negative stain. Observe the stain and describe your results.
5. Wash your slides. Wipe the oil off your microscope and return it.

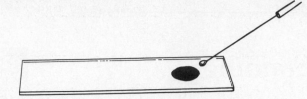

(a) Place a small drop of nigrosine near one end of a slide. Mix a loopful of broth culture in the drop. When the organisms are taken from a solid medium, mix a loopful of water in the nigrosine.

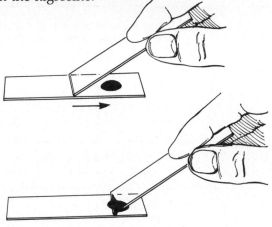

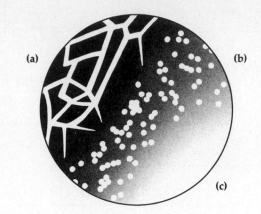

Figure 4.2 —————

A negative stain viewed under a microscope. **(a)** This part of the smear is too heavy. **(b)** Colorless cells are visible here. **(c)** Too little stain is in this area of the smear.

(b) Gently draw a second slide across the surface of the first until it contacts the drop. The drop will spread across the edge of the top slide.

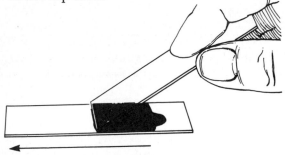

(c) Push the top slide to the left along the entire surface of the bottom slide.

(d) Let the smear air dry.

Figure 4.1 —————

Preparing a negative stain.

EXERCISE 4

Negative Staining

Name _ANNABELLE G. BUENDIA_

Date _JUNE 8, 1993_

Lab Section _____

Purpose _Explain the application + mechanism of the negative stain technique. Prepare a negative stain._

Data

Appearance	*Bacillus subtilis*	*Staphylococcus epidermidis*	Tooth and gum scraping
Sketch a few bacteria (1000 ×)			
Morphology and arrangement			

Questions

1. Which cell is a rod? _____ How does its appearance differ from the rod used in Exercise 3? _____

2. What microscopic technique gives a field similar in appearance to that seen in the negative stain? _____

3. Why is the size more accurate in a negative stain than in a simple stain? _____

4. Could any dye be used in place of nigrosine for negative staining? _____

   What types of dyes are used for negative staining? _____

EXERCISE 5

Gram Staining

EXERCISE

5

Objectives

After completing this exercise you should be able to:

1. Explain the rationale and procedure for the Gram stain.
2. Perform and interpret Gram stains.

Background

The Gram stain is a very useful stain for identifying and classifying bacteria. The **Gram stain** is a differential stain that allows you to classify bacteria as either gram-positive or gram-negative. The Gram staining technique was discovered by Hans Christian Gram in 1884, when he attempted to stain cells and found that some lost their color when excess stain was washed off.

The staining technique consists of the following steps:

1. Apply **primary stain** (crystal violet). All bacteria are stained purple by this basic dye.
2. Apply **mordant** (Gram's iodine). The iodine combines with the crystal violet in the cell to form a crystal violet–iodine complex (CV–I).
3. Apply **decolorizing** (ethyl alcohol or ethyl alcohol–acetone). The primary stain is washed out (decolorized) of some bacteria, while others are unaffected.
4. Apply **secondary stain** or **counterstain** (safranin). This basic dye stains the decolorized bacteria red.

This is the most common technique, although variations have been developed. The most important determining factor in the procedure is that bacteria differ in their *rate* of decolorization. Those that decolorize easily are referred to as **gram-negative,** whereas those that retain the primary stain are called **gram-positive.**

Bacteria stain differently because of chemical and physical differences in their cell walls. Gram-positive cell walls consist of many layers of peptidoglycan. The CV–I is larger than the crystal violet or iodine molecules that initially entered the cell and cannot pass through the thick peptidoglycan. In gram-negative cells, the alcohol dissolves the outer lipopolysaccharide layer, and the CV–I washes out through the thin layer of peptidoglycan.

The Gram stain is most consistent when done on young cultures of bacteria (less than 24 hours old). Because Gram staining is usually the first step in identifying bacteria, the procedure should be memorized.

Materials

Gram staining reagents:
    Crystal violet
    Gram's iodine
    Ethyl alcohol
    Safranin

Wash bottle of distilled water

Slides (3)

Cultures

Staphylococcus epidermidis

Escherichia coli

Bacillus subtilis

Techniques Required

Exercises 1 and 3.

Procedure

1. Prepare a smear (Exercise 3). Follow procedure a or b.
    a. Clean three slides well and make a circle with a marker on each slide. Label each slide for one of the cultures.
    b. Clean one slide well and make three circles with a marker on the slide. Label each circle for one of the cultures.
2. Prepare a Gram stain. Use a clothespin or slide rack to hold the slide.
    a. Cover the smear with crystal violet and leave for 30 seconds (Figure 5.1*a*).
    b. Wash the slide carefully with distilled water from a wash bottle. Do not squirt water directly onto the smear (Figure 5.1*b*).
    c. Without drying, cover the smear with Gram's iodine for 30 seconds (Figure 5.1*c*).

33

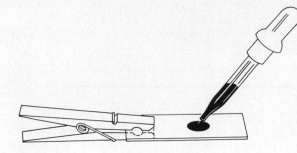

(a) Cover the smear with crystal violet for 30 seconds.

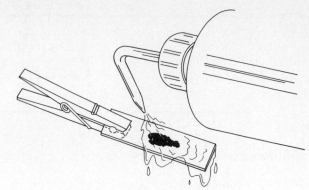

(b) Gently wash off the crystal violet with water by spraying the water so it runs through the smear.

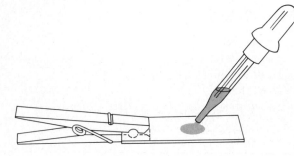

(c) Cover the smear with Gram's iodine for 30 seconds.

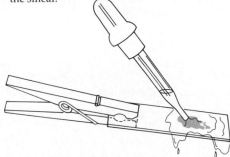

(d) Wash off the iodine with alcohol.

(e) Gently wash the smear with water.

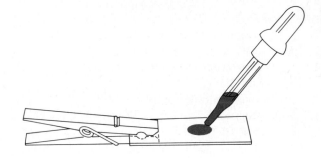

(f) Cover the smear with safranin for 30 seconds.

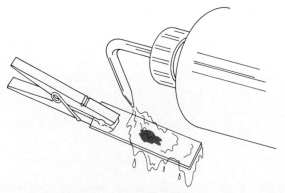

(g) Wash with water and blot dry.

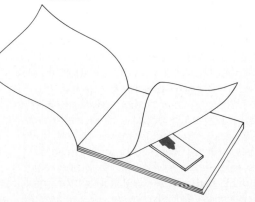

Figure 5.1 _____

The Gram stain.

d. Without washing, decolorize with 95% ethyl alcohol (Figure 5.1*d*). Let the alcohol run through the smear until no large amounts of purple wash out (usually a few seconds). The degree of alcohol decolorizing depends on the thickness of the smear. This is a critical step. *Do not over decolorize.* However, experience is the only way you will be able to determine how long to decolorize.

e. Immediately wash gently with distilled water (Figure 5.1*e*). Why? _____

f. Add safranin for 30 seconds (Figure 5.1*f*).

g. Wash with distilled water and blot the slide dry with a paper towel or absorbent paper (Figure 5.1*g*).

3. Examine the stained slide microscopically using the low, high-dry, and oil immersion objectives. Put the oil directly on the smear. Record your observations. (See color plates I.1 and I.2.) Do they agree with those given in your textbook? _____ If not, try to determine why. Some common sources of Gram staining errors are:

a. The loop was too hot.

b. Excessive heat was applied during heat fixing.

c. Decolorizing agent (alcohol) was left on the smear too long.

4. Stain the remaining two slides.

5. Design your own differential stain using different reagents but the same general scheme. Does it work? _____

EXERCISE 5

Gram Staining

Name _____

Date _____

Lab Section _____

Purpose _____

_____

_____

Data

Appearance	Staphylococcus epidermidis	Bacillus subtilis	Escherichia coli
Sketch a few bacteria (1000 ×)			
Morphology, arrangement, relative size			
Color			
Gram reaction			

Questions

1. Can iodine be added before the primary stain in a Gram stain? _____

_____

_____

2. If you Gram stained human cells, what would happen? _____

_____

_____

3. Why will old gram-positive cells stain gram-negative? _____

_____

_____

4. Did your results agree with the information in your textbook? _____ If not, why not?

_____

_____

5. What other properties can be associated with the chemical differences of the cell wall? For example, can dye

sensitivity be so associated? _____

_____

_____

6. List the steps of the Gram staining procedure in order (omit washings) and fill in the color of gram-positive cells
and gram-negative cells after each step.

Step	Chemical	Appearance	
		Gram-positive cells	Gram-negative cells
1			
2			
3			
4			

7. Which step can be omitted without affecting determination of the Gram reaction? _____

EXERCISE 6

Acid-Fast Staining

Objectives

After completing this exercise you should be able to:

1. Apply the acid-fast procedure.
2. Explain what is occurring during the acid-fast staining procedure.
3. Perform and interpret an acid-fast stain.

Background

The **acid-fast stain** is a differential stain. In 1882 Paul Ehrlich discovered that *Mycobacterium tuberculosis* (the causative agent of tuberculosis) retained the primary stain even after washing with an acid-alcohol mixture. (We hope you can appreciate the phenomenal strides that were made in microbiology in the 1880s. Most of the staining and culturing techniques used today originated during that time.) Most bacteria are decolorized by acid-alcohol, with only the families Mycobacteriaceae and Nocardiaceae of the order Actinomycetales *(Bergey's Manual, Volume 4*)* being acid-fast. The acid-fast technique has great value as a diagnostic procedure because both *Mycobacterium* and *Nocardia* contain **pathogenic** (disease-causing) species.

The cell walls of acid-fast organisms contain a wax-like lipid called **mycolic acid,** which renders the cell wall impermeable to most stains. The cell wall is so impermeable, in fact, that a clinical specimen is usually treated with strong sodium hydroxide to remove debris and contaminating bacteria prior to culturing mycobacteria.

Today, the technique developed by Franz Ziehl and Friedrich Neelsen is the most widely used acid-fast stain. In the **Ziehl–Neelsen procedure,** the smear is flooded with carbolfuchsin (containing 5% phenol), which has a high affinity for a chemical component of the bacterial cell. The smear is heated to facilitate penetration of the stain into the bacteria. The stained smears are washed with an acid-alcohol mixture that easily decolorizes most bacteria except the acid-fast microbes. Methylene blue is then used as a counterstain to enable you to observe the non–acid-fast organisms.

The mechanism of the acid-fast stain is probably the result of the relative solubility of carbolfuchsin and the impermeability of the cell wall. Fuchsin is more soluble in carbolic acid (phenol) than in water, and carbolic acid solubilizes more easily in lipids than in acid-alcohol. Therefore, carbolfuchsin has a higher affinity for lipids than acid-alcohol and will remain with the cell wall when washed with acid-alcohol.

Materials

Acid-fast staining reagents:
    Ziehl's carbolfuchsin
    Acid-alcohol
    Methylene blue

Wash bottle of distilled water

Slides (2)

Boiling water bath

Cultures

Mycobacterium phlei

Escherichia coli

Demonstration Slides

Acid-fast sputum slides

Techniques Required

Exercises 1 and 3.

Procedure

1. Prepare and heat fix a smear (Exercise 3) of one culture.
2. Cover the smear with a small piece of absorbent paper (Figure 6.1*a*). The paper will reduce evaporation of the stain.
3. Add a copious amount of Ziehl's carbolfuchsin (Figure 6.1*b*).
4. Heat by setting the slide on a boiling water bath (Figure 6.1*c*). Steam for 5 minutes, adding more stain as it is needed.
5. Discard the paper; *do not put it in the sink.* Wash the slide well with distilled water, then decolorize for 15 to 20 seconds with acid-alcohol. Wash with distilled water.

---

*J. G. Holt, cd. *Bergey's Manual of Systematic Bacteriology*, 1st ed. Baltimore: Williams & Wilkins, 1984–1989. See also *Bergey's Manual of Determinative Bacteriology*, 9th ed. Baltimore: Williams & Wilkins, 1992.

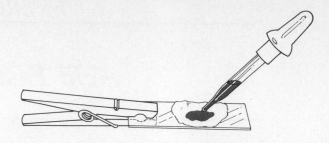

(a) Place a piece of absorbent paper over the smear.

(b) Cover the paper with Ziel's carbolfuchsin.

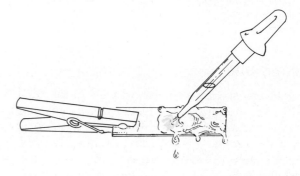

(c) After steaming for 5 minutes, wash the smear with acid-alcohol.

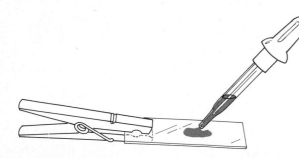

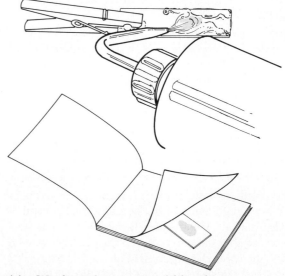

(d) Cover the smear with methylene blue for 30 seconds.

(e) Wash with water and blot dry.

Figure 6.1 _____

Preparing an acid-fast stain.

6. Counterstain for about 30 seconds with methylene blue (Figure 6.1*d*).
7. Wash with distilled water and blot dry (Figure 6.1*e*).

8. Prepare an acid-fast stain of the other culture.
9. Examine the acid-fast-stained slides microscopically and record your observations.
10. Observe the demonstration slides.

EXERCISE 6

Acid-Fast Staining

Name _____

Date _____

Lab Section _____

Purpose _____

_____

_____

Data

Appearance	*Mycobacterium phlei*	*Escherichia coli*	Demonstration slides	
			#1	#2
Sketch a few bacteria (1000 ×)				
Morphology and color				
Acid-fast reaction				

Questions

1. What are the large stained areas on the sputum slide? _____

   _____

   _____

2. How do the acid-fast properties relate to the Gram stain? _____

   _____

3. Assuming you could stain any cell, would an acid-fast organism be gram-positive or gram-negative? Explain.

   _____

   _____

4. What diseases are diagnosed using the acid-fast procedure? _____

_____

_____

5. What is phenol (carbolic acid), and what is its *usual* application? _____

_____

_____

Structural Stains (Endospore, Capsule, and Flagella)

Objectives

After completing this exercise you should be able to:

1. Prepare and interpret endospore, capsule, and flagella stains.
2. Recognize the different types of flagellar arrangements.

Background

Structural stains can be used to identify and study the structure of bacteria. Currently, most of the fine structural details are examined using an electron microscope, but historically, staining techniques have given much insight into bacterial fine structure. We will examine a few structural stains that are still useful today. These stains are used to observe endospores, capsules, and flagella.

Endospores

Endospores are formed by members of six genera included in *Bergey's Manual,** endospore-forming gram-positive rods and cocci. *Bacillus* and *Clostridium* are the most familiar genera. Endospores are called "resting bodies" because they do not metabolize and are resistant to heating, various chemicals, and many harsh environmental conditions. Endospores are not for reproduction; they are formed when essential nutrients or water are not available. Once an endospore forms in a cell, the cell will disintegrate (Figure 7.1). Endospores can remain dormant for long periods of time. However, an endospore may return to its vegetative or growing state.

*J. G. Holt, ed. *Bergey's Manual of Systematic Bacteriology*, 1st ed. Baltimore: Williams & Wilkins, 1984–1989. See also *Bergey's Manual of Determinative Bacteriology*, 9th ed. Baltimore: Williams & Wilkins, 1992.

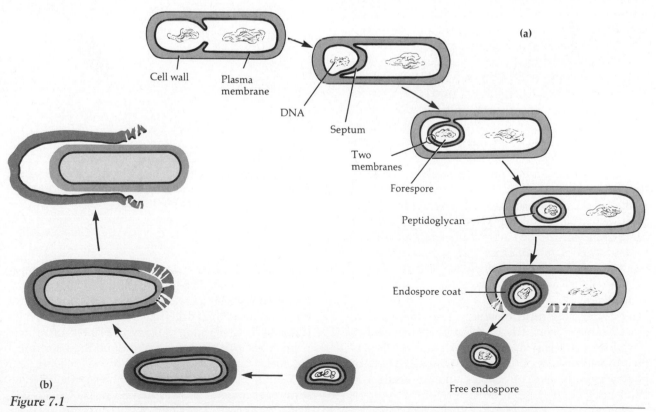

Figure 7.1

(a) Sporogenesis, the process of endospore formation. **(b)** Germination of an endospore to a vegetative cell.

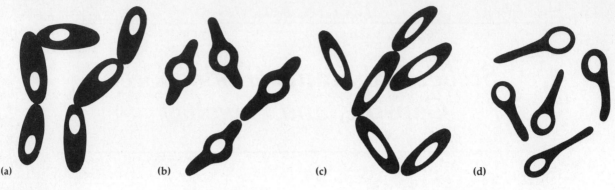

Figure 7.2 _____

Some examples of bacterial endospores. **(a)** Subterminal endospores (*Bacillus macerans*). **(b)** Central, swollen endospores (*Clostridium perfringens*). **(c)** Central endospores (*Bacillus polymyxa*). **(d)** Terminal, swollen endospores (*Clostridium tetani*).

Taxonomically, it is very helpful to know whether a bacterium is an endospore former and also the position of the endospores (Figure 7.2). Endospores are impermeable to most stains, so heat is usually applied to drive the stain into the endospore. Once stained, the endospores do not readily decolorize. We will use the **Schaeffer–Fulton** endospore stain.

Capsules

Many bacteria secrete chemicals that adhere to their surfaces, forming a viscous coat. This structure is called a **capsule** when it is round or oval in shape, and a **slime layer** when it is irregularly shaped and loosely bound to the bacterium. The ability to form capsules is genetically determined, but the size is influenced by the medium on which the bacterium is growing. Most capsules are composed of polysaccharides, which are water-soluble and uncharged. Because of the capsule's nonionic nature, simple stains will not adhere to it. Most capsule staining techniques stain the bacteria and the background, leaving the capsules unstained — essentially, a "negative" capsule stain.

Capsules have a very important role in the **virulence** (disease-causing ability) of some bacteria. For example, when bacteria such as *Streptococcus pneumoniae* have a capsule, the body's white blood cells cannot phagocytize the bacteria efficiently, and disease occurs. When *S. pneumoniae* lack a capsule, they are easily engulfed and are not virulent.

Flagella

Many bacteria are **motile,** which means they have the ability to move from one position to another in a directed manner. Most motile bacteria possess flagella, but other forms of motility occur. Myxobacteria exhibit gliding motion, and spirochetes undulate using axial filaments.

Flagella, the most common means of motility, are thin proteinaceous structures that originate in the cytoplasm and project out from the cell wall. They are

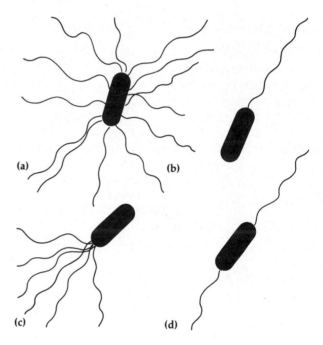

Figure 7.3 _____

Flagellar arrangements. **(a)** Peritrichous flagella. Types of polar flagella: **(b)** monotrichous flagella, **(c)** lophotrichous flagella, and **(d)** amphitrichous flagella.

very fragile and are not visible with a light microscope. They can be stained after carefully coating them using a mordant, which increases their diameter. The presence and location of flagella are helpful in the identification and classification of bacteria. Flagella are of two main types: **peritrichous** (all around the bacterium) and **polar** (at one or both ends of the cell) (Figure 7.3).

Motility may be determined by observing hanging-drop preparations of unstained bacteria (Exercise 2), flagella stains, or inoculation of soft (or semisolid) agar (Exercise 10). If time does not permit doing flagella stains, observe the demonstration slides.

Materials

Slides

Cover slip

Paper towels

Wash bottle of distilled water

Forceps

Scalpel

Endospore stain reagents: malachite green and safranin

Capsule stain reagents: Congo red, acid-alcohol, and acid fuchsin

Flagella stain reagents: flagella mordant and Ziehl's carbolfuchsin

Cultures (as needed)

Endospore stain

Bacillus megaterium (24-hour)

Bacillus subtilis (24-hour)

Bacillus subtilis (72-hour)

Capsule stain

Streptococcus salivarius

Enterobacter aerogenes

Flagella stain

Proteus vulgaris (18-hour)

Demonstration Slides

Endospore stain

Capsule stain

Flagella stain

Techniques Required

Exercises 1, 3, and 4.

Procedure

Endospore Stain (Figure 7.4)

Be careful. The malachite green has a messy habit of ending up everywhere. But most likely you will end up with green fingers no matter how careful you are.

1. Clean three slides. Make smears of the three *Bacillus* cultures, air dry, and heat fix (Exercise 3).
2. Tear out small pieces of paper towel and place on the slides, to reduce evaporation of the stain. The paper should be smaller than the slide (Figure 7.4*a*).

3. Flood the smear and paper with malachite green; steam for 5 minutes. Add more stain as needed. *Keep it wet* (Figure 7.4*b*). What is the purpose of the paper? _____
4. Remove the towel and discard carefully. *Do not put it in the sink.* Wash the stained smears well with distilled water (Figure 7.4*c*).
5. Counterstain with safranin for 30 seconds (Figure 7.4*e*).
6. Wash with distilled water and blot dry (Figure 7.4*f*).
7. Examine microscopically and record your observations.
8. Observe the demonstration slides of bacterial endospores.

Capsule Stain

1. Prepare a thick smear of bacteria in a loopful of Congo red. Let the smear air dry. What color is the smear? _____
2. Fix the smear with acid-alcohol for 15 seconds. What color is the smear? _____
3. Wash with distilled water and cover the smear with acid fuchsin for 1 minute.
4. Wash with distilled water, blot dry, and examine microscopically. The bacteria will stain red, and the capsules will be colorless against a dark blue background. Record your observations.
5. Observe the demonstration slides of capsule stains.

Flagella Stain

1. Flagella stains require special precautions to avoid damaging the flagella. Scrupulously clean slides are essential, and the culture must be handled carefully to prevent flagella from coming off the cells.
2. Without touching the bacterial culture, use a scalpel and forceps to cut out a piece of agar on which *Proteus* is growing.

> ☣ **Do not touch the culture with your hands.**

Gently place the agar, culture-side down, on a clean glass slide. Then carefully remove the agar with the forceps. Place the piece of agar in a Petri plate.
3. Allow the organisms adhering to the slide to air dry. *Do not heat fix.*
4. Cover the slide with flagella mordant and allow it to stand for 10 minutes.
5. Gently rinse off the stain with distilled water.
6. Cover the slide with Ziehl's carbolfuchsin for 5 minutes. Rinse gently with distilled water. What color should the flagella be? _____ The cells? _____
7. Allow the rinsed smear to air dry *(do not blot)* and examine microscopically for flagella.
8. Observe the demonstration slides illustrating various flagellar arrangements. (See color plate I.3.)

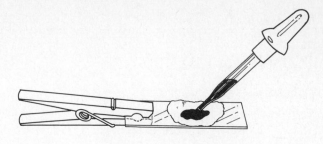

(a)   Place a piece of absorbent paper over the smear.

(b)   Cover the paper with malachite green.

(c)   After steaming for 5 minutes, wash the smear with water.

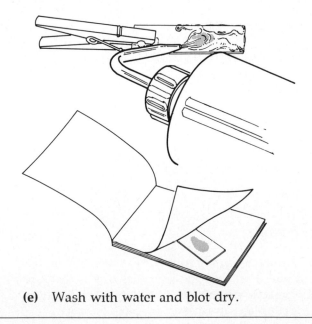

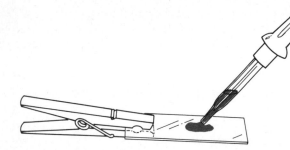

(d)   Cover the smear with safranin for 30 seconds.

(e)   Wash with water and blot dry.

Figure 7.4 _____
The endospore stain.

EXERCISE 7

Structural Stains (Endospore, Capsule, and Flagella)

Name _____

Date _____

Lab Section _____

Purpose _____

_____

_____

Data

Endospores

Sketch your results and label each diagram as to color. Label the vegetative cells and endospores.

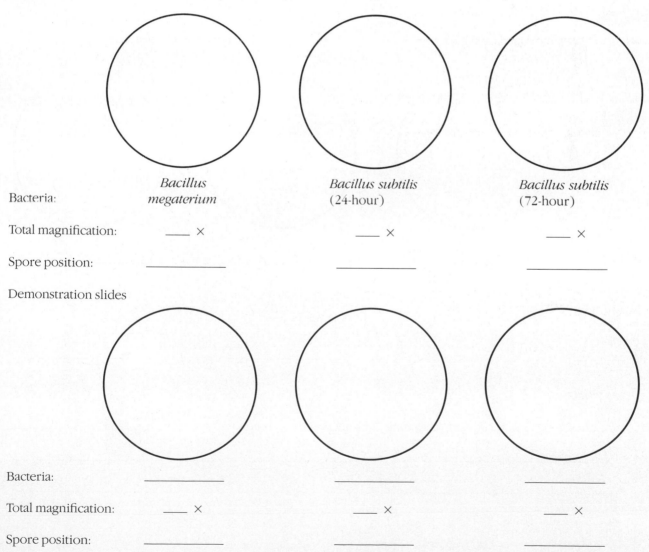

Bacteria: *Bacillus megaterium* | *Bacillus subtilis* (24-hour) | *Bacillus subtilis* (72-hour)

Total magnification: ____ × ____ × ____ ×

Spore position: _____ _____ _____

Demonstration slides

Bacteria: _____ _____ _____

Total magnification: ____ × ____ × ____ ×

Spore position: _____ _____ _____

Capsules

Sketch and label the capsules and bacterial cells.   Demonstration slide

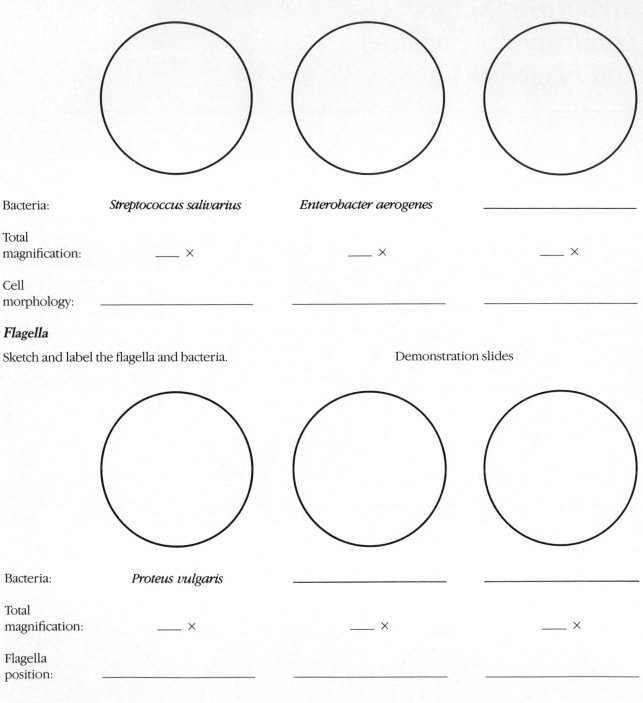

Bacteria: *Streptococcus salivarius*   *Enterobacter aerogenes*   _____

Total
magnification: ____ ×   ____ ×   ____ ×

Cell
morphology: _____   _____   _____

Flagella

Sketch and label the flagella and bacteria.   Demonstration slides

Bacteria: *Proteus vulgaris*   _____   _____

Total
magnification: ____ ×   ____ ×   ____ ×

Flagella
position: _____   _____   _____

Questions

1. You can see endospores by simple staining. Why not use this technique? _____

_____

_____

2. How would an endospore stain of *Mycobacterium* appear? _____
_____
_____

3. What are the Gram reactions of *Clostridium* and *Bacillus*? _____
_____

4. What type of culture medium would increase the size of a bacterial capsule? _____
_____
_____

5. How might a capsule contribute to pathogenicity? _____
   Flagella? _____

6. Of what morphology are most bacteria possessing flagella? _____
   Which morphology usually does not have flagella? _____

7. What prevents the cell from appearing green in the finished endospore stain? _____
_____
_____

8. Of what advantage to *Clostridium* is an endospore? _____
_____
_____

9. Write a definition for each of the following flagellar arrangements:

   a. Monotrichous _____

   b. Lophotrichous _____

   c. Amphitrichous _____

   d. Peritrichous _____

Morphologic Unknown

*Dishonesty is knowing but ignoring the fact
that the data are contradictory. Stupidity is
not recognizing the contradictions.*

ANONYMOUS

Objective

After completing this exercise you should be able to:

1. Identify the morphology and staining characteristics of an unknown organism.

Background

To identify a bacterium, it is very important to determine morphology, arrangement, Gram reaction, and structural details. You will be given an unknown culture of bacteria. Determine its morphologic and structural characteristics. The culture contains one species (rod or cocci) and is less than 24 hours old.

Materials

Staining reagents

Culture

24-hour unknown slant culture of bacteria # _____

Techniques Required

Exercises 1, 2, and 3–7.

Procedure

1. Record the number of your unknown.
2. Determine the morphology, Gram reaction and arrangement of your unknown. Perform a Gram stain and, if needed, an endospore stain, acid-fast stain, flagella stain, hanging-drop technique, or capsule stain. When are the latter needed? _____
3. Tabulate your results in the Laboratory Report.

EXERCISE 8

Morphologic Unknown

Name _____

Date _____

Lab Section _____

Results

Write *not done* by any category that does not apply.

Unknown # _____

Sketch of the Gram stain

_____ ×

Gram reaction: _____

Morphology: _____

Predominant arrangement: _____

Endospores present? _____

Sketch if present

Capsules present? _____

Sketch if present

Motile? _____

Motility determined by: _____

Acid-fast? _____

PART 3

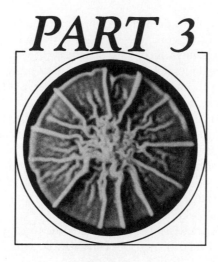

Cultivation of Bacteria

EXERCISES

9 **Microbes in the Environment**
10 **Transfer of Bacteria: Aseptic Technique**

11 **Isolation of Bacteria by Dilution Techniques**
12 **Special Media for Isolating Bacteria**

In the previous exercises you saw that bacteria are grown on culture media in a laboratory. Bacteria must be cultured in order to characterize and identify them. This statement from the first edition of *Bergey's Manual* summarizes the need for cultivating bacteria:

> *The earlier writers classified the bacteria solely on their morphologic characters. A more detailed classification was not possible because the biologic characters of so few of the bacteria had been determined. With the accumulation of knowledge of the biologic characters of many bacteria it was realized that it is just as incorrect to group all rods under a single genus as to group all quadruped animals under one genus.* *

*Society of American Bacteriologists (D. H. Bergey, Chairperson). *Bergey's Manual of Determinative Bacteriology,* 1st ed. Baltimore: Williams & Wilkins, 1923, p. 1.

Before attempting to culture bacteria, we must consider their nutritional requirements. Bacteria require sources of energy, carbon, nitrogen, minerals, and growth factors.

A culture medium whose exact chemical composition is known is called a **chemically defined medium.** An example of a chemically defined medium is shown in Table 1.

Most bacteria are routinely grown on **complex media,** that is, media for which the exact chemical composition varies slightly from batch to batch. Organic carbon, energy, and nitrogen sources are usually supplied by protein in the form of meat extracts and partially digested proteins called *peptones*. **Nutrient broth** is a commonly used liquid complex medium. When agar is added, it becomes a solid medium called **nutrient agar** (Table 2).

The first bacterial cultures were grown in *broth* (liquid) media such as infusions and blood. Robert Koch attempted to culture bacteria on potato slices when he realized the need for *solid media*. When some bacteria wouldn't grow on a potato, he added gelatin to the liquid, but the gelatin liquefied under standard incubation conditions. The wife of Walter Hess, a colleague of Koch's, suggested the use of agar as a solidifying agent. Today, agar is the most commonly used solidifying agent in culture media.

Agar, an extract from marine red algae, has some unique properties that make it useful in culture media. Few microbes can degrade agar, so it remains solid during microbial growth. It liquefies at 100°C and remains in a liquid state until cooled to 40°C. Once the agar has solidified, it can be incubated at temperatures up to 100°C and remain solid.

The exercises in Part 3 emphasize the use of **aseptic technique,** the prevention of unwanted microorganisms in laboratory and medical procedures.

Table 1
Glucose-minimal Salts Broth

Ingredient	Amount/100 ml
Glucose	0.5 g
Sodium chloride (NaCl)	0.5 g
Ammonium dihydrogen phosphate ($NH_4H_2PO_4$)	0.1 g
Dipotassium phosphate (K_2HPO_4)	0.1 g
Magnesium sulfate ($MgSO_4$)	0.02 g
Distilled water	100 ml

Table 2
Nutrient Agar

Ingredient	Amount/100 ml
Peptone	0.5 g
Beef extract	0.3 g
Sodium chloride (NaCl)	0.8 g
Agar	1.5 g
Distilled water	100 ml

Microbes in the Environment

*Whatever is worth doing at all
is worth doing well.*

PHILIP DORMER STANHOPE,
EARL OF CHESTERFIELD

Objectives

After completing this exercise you should be able to:

1. Describe colony morphology using accepted descriptive terms.
2. Compare bacterial growth on solid and liquid culture media.

Background

Microbes are everywhere; they are found in the water we drink, the air we breathe, and the earth we walk on. They live in and on our bodies. Microbes occupy ecological niches on all forms of life and in most environments. In most situations, the ubiquitous microorganisms are harmless. However, in microbiology, work must be done carefully to avoid contaminating sterile media and materials with these microbes.

In this exercise, we will attempt to culture (grow) some microbes. Culture media can be prepared in various forms, depending on the desired use. **Petri plates** containing solid media provide a large surface area for examination of colonies. The microbes will be **inoculated,** or intentionally introduced, onto nutrient agar and into nutrient broth. The bacteria that are inoculated into culture media increase in number during an **incubation period.** After suitable incubation, liquid media become **turbid,** or cloudy, due to bacterial growth. On solid media, colonies will be visible to the naked eye. A **colony** is a population of cells that arises from a single bacterial cell. Although many species of bacteria give rise to similar appearing colonies, each different appearing colony is usually a different species.

Materials

Petri plates containing nutrient agar (4)

Tube containing nutrient broth

Tube containing sterile cotton swabs

Tube containing sterile water

Techniques Required

None.

Procedure

First Period

1. Design your own experiment. The purpose is to sample your environment and your body. Use your imagination. Here are some suggestions:
   a. You may use the lab, a washroom, or any place on campus for the environment.
   b. One nutrient agar plate might be left open to the air for 30 to 60 minutes.
   c. Inoculate a plate from an environmental surface such as the floor or workbench by wetting a cotton swab in sterile water, swabbing the environmental surface, and then swabbing the surface of the agar. Why is the swab first moistened in sterile water? _____

   _____

   After inoculation, the swab should be discarded in the container of disinfectant.
2. Inoculate two plates from the environment. Inoculate one nutrient broth tube using a swab as described in step 1c. After swabbing the agar surface, place the swab in the broth and leave it there during incubation.
3. The plates and tube should be incubated at the approximate temperature of the environment sampled.
4. Inoculate two plates from your body. You could:
   a. Place a hair on the agar.
   b. Obtain an inoculum by swabbing (see step 1c) part of your body.
   c. Touch the plate.
5. Incubate bacteria from your body at or close to

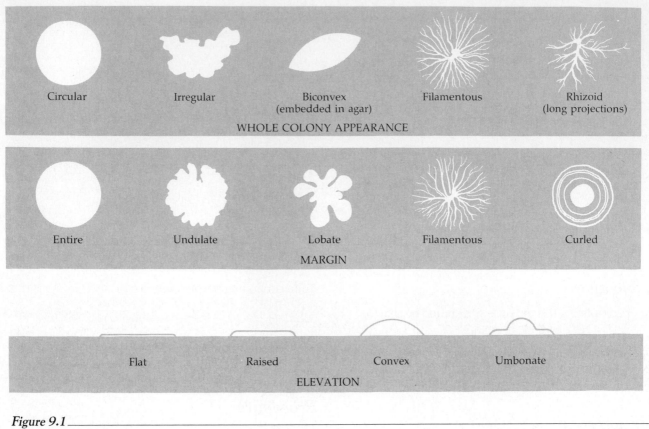

Figure 9.1

Colony descriptions.

your body temperature. What is human body temperature? _____°C

6. Incubate all plates, inverted so water will condense in the lid instead of on the surface of the agar. Why is condensation on the agar undesirable? _____

_____

7. Incubate all inoculated media until the next laboratory period.

Second Period

1. Observe and describe the resulting growth on the plates. Note each different appearing colony, and describe the colony morphology using the charac-

teristics given in Figure 9.1. Determine the approximate number of each type of colony. When many identical colonies are present, record TMTC (too many to count) as the number of colonies.

2. Describe the appearance of the nutrient broth. Is it uniformly cloudy or turbid? Look for clumps of microbial cells, called **flocculant.** Is there a membrane, or **pellicle,** across the surface of the broth? See whether microbial cells have settled on the bottom of the tube forming a **sediment.**

3. Discard the plates properly and save the tube of broth for Exercise 11.

EXERCISE 9

Microbes in the Environment

Name _____

Date _____

Lab Section _____

Purpose _____

_____

_____

Data

Fill in the following table with descriptions of the bacterial colonies. Use a separate line for each different appearing colony.

	Colony Description					
	Diameter	Appearance	Margin	Elevation	Color	Number of this type
Area sampled _____ Incubated at _____°C for _____ hrs.						
Area sampled _____ Incubated at _____°C for _____ hrs.						
Area sampled _____ Incubated at _____°C for _____ hrs.						

	Colony Description					
	Diameter	Appearance	Margin	Elevation	Color	Number of This Type
Area sampled _____ Incubated at _____°C for _____ hrs.						

Description of the nutrient broth:

Area sampled: _____ . Incubated at _____ °C for _____ hours.

Is it turbid? _____

Is a flocculant present? _____

Is a sediment present? _____

Is a pellicle present? _____

Has the color changed? _____

Questions

1. How can you tell whether or not there is bacterial growth in nutrient broth? _____

2. What is the minimum number of different bacteria present on one of your plates? _____

   How do you know? _____

3. Did all the organisms living in or on the environments sampled grow on your nutrient agar? _____

   Briefly explain. _____

4. Of what advantage is a solid medium over a liquid medium? _____

5. What is the value of Petri plates in microbiology? _____

Transfer of Bacteria: Aseptic Technique

*Study without thinking is worthless;
thinking without study is dangerous.*

CONFUCIUS

Objectives

After completing this exercise you should be able to:

1. Provide the rationale for aseptic technique.
2. Differentiate among the following: broth culture, agar slant, and agar deep.
3. Aseptically transfer bacteria from one form of culture medium to another.

Background

In the laboratory, bacteria must be cultured in order to facilitate identification and to examine their growth and metabolism. Bacteria are **inoculated,** or introduced, into various forms of culture media in order to keep them alive and to study their growth. Inoculations must be done without introducing unwanted microbes, or **contaminants,** into the media. **Aseptic technique** is used in microbiology to exclude contaminants.

All culture media are **sterilized,** or rendered free of all life, prior to use. Sterilization is usually accomplished using an autoclave. Containers of culture media, such as test tubes or Petri plates, should not be opened until you are ready to work with them, and even then, they should not be left open.

Broth cultures provide large numbers of bacteria in a small space and are easily transported. **Agar slants** are test tubes containing solid culture media that were left at an angle while the agar solidified. Agar slants, like Petri plates, provide a solid growth surface, but slants are easier to store and transport than Petri plates. Agar is allowed to solidify in the bottom of a test tube to make an **agar deep.** Deeps are often used to grow bacteria that prefer less oxygen than is present on the surface of the medium. Semisolid agar deeps containing 0.5%–0.7% agar instead of the usual 1.5% agar can be used to determine whether a bacterium is motile (Exercise 7). Motile bacteria will move away from the point of inoculation, giving an inverted "Christmas tree" appearance.

Transfer and inoculation are usually performed with a sterile, heat-resistant, noncorroding nichrome wire attached to an insulated handle. When the end of the wire is bent into a loop, it is called an **inoculating loop;** when straight, it is an **inoculating needle** (Figure 10.1). For special purposes, cultures may also be transferred with sterile cotton swabs, pipettes, glass rods, or syringes. These techniques will be introduced in later exercises.

Whether an inoculating loop or needle is used depends on the form of the medium; after completing this exercise, you will be able to decide which instrument is to be used.

Materials

Tubes containing nutrient broth (3)

Tubes containing nutrient agar slants (3)

Tubes containing nutrient semisolid agar deeps (3)

Inoculating loop

Inoculating needle

Test tube rack

Gram staining reagents

Cultures

Streptococcus lactis broth

Pseudomonas aeruginosa broth

Proteus vulgaris slant

Techniques Required

Exercise 5.

Procedure

1. Work with only one of the bacterial cultures at a time, to prevent any mixups or cross-contamina-

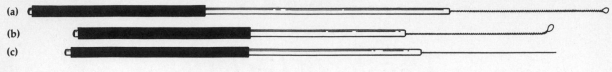

Figure 10.1

(a) An inoculating loop. **(b)** A variation of the inoculating loop in which the loop is bent at a 45° angle. **(c)** An inoculating needle.

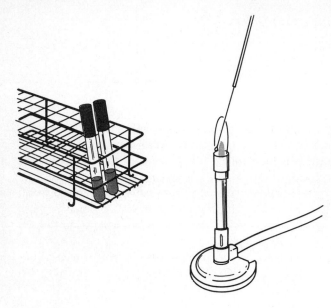

(a)   Sterilize the loop by holding the wire in the flame until it is red hot.

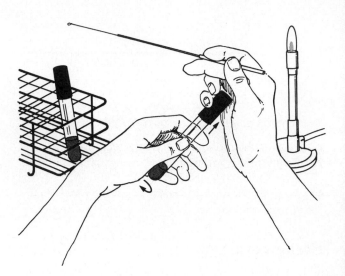

(b)   While holding the sterile loop and the bacterial culture, remove the cap as shown.

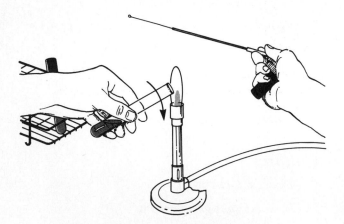

(c)   Briefly heat the mouth of the tube in the flame before inserting the loop for an inoculum.

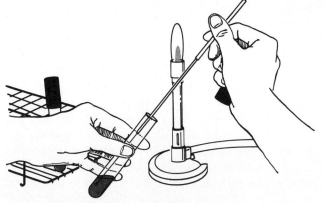

(d)   Get a loopful of culture, heat the mouth of the tube, and replace the cap.

Figure 10.2

Inoculating procedures.

tion. Begin with one of the broth cultures, and gently tap the bottom of it to resuspend the sediment.

2. To inoculate nutrient broth, hold the inoculating loop in your dominant hand and one of the broth cultures of bacteria in the other hand.

   a. Sterilize the loop by holding the wire in a Bunsen burner flame (Figure 10.2*a*). Heat to redness. Why? _____

   b. Holding the loop like a pencil, curl the little finger of the same hand around the cap of the broth culture. Gently pull the cap off the tube while turning the culture tube (Figure 10.2*b*). If cotton stoppers are used, simply grasp the stopper with your finger. Do not set the cap down. Why not? _____

   c. Holding the tube at an angle, pass the mouth of the tube through the flame (Figure 10.2*c*). What is the purpose of flaming the mouth of the tube? _____
   Always hold culture tubes and uninoculated tubes at an angle to minimize the amount of dust that could fall into them.

   d. Immerse the sterilized, cooled loop into the broth culture to obtain a loopful of culture (Figure 10.2*d*). Why must the loop be cooled first? _____
   Remove the loop, and while holding the loop, flame the mouth of the tube and recap by turning the tube into the cap. Place the tube in your rack.

   e. Remove the cap from a tube of sterile nutrient broth as previously described, and flame the mouth of the tube. Immerse the inoculating loop into the sterile broth. Flame the mouth of the tube and replace the cap. Return the tube to the test tube rack.

   f. Reflame the loop until it is red and let it cool. Some prefer to hold several tubes in their hands at once (Figure 10.3). *Do not* attempt holding and transferring between multiple tubes until you have mastered aseptic transfer techniques.

   g. Label the inoculated tube with your name, the date, and your lab section.

3. Obtain a nutrient agar slant. Repeat steps 2a–2d, and inoculate the slant by moving the loop gently across the agar surface from the bottom of the slant to the top, being careful not to gouge the agar (Figure 10.4). Flame the mouth of the tube and replace the cap. Flame your loop and let it cool.

4. Obtain a nutrient agar semisolid deep, and using your inoculating needle, repeat steps 2a–2d. Inoculate the semisolid agar deep by plunging the needle straight down the middle of the deep, then pull out through the same stab, as shown in Figure 10.5. Flame the mouth of the tube and replace the cap. Flame your needle and let it cool.

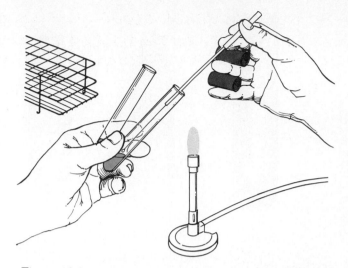

Figure 10.3 _____
Experienced laboratory technicians can transfer cultures aseptically, holding multiple test tubes.

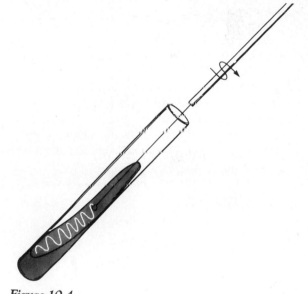

Figure 10.4 _____
Inoculate a slant by streaking back and forth across the surface of the agar.

Figure 10.5 _____
Inoculate an agar deep by stabbing into the agar with a needle.

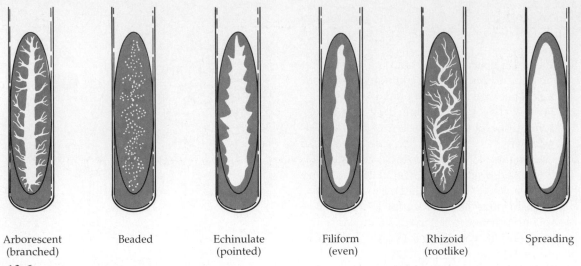

| Arborescent (branched) | Beaded | Echinulate (pointed) | Filiform (even) | Rhizoid (rootlike) | Spreading |

Figure 10.6

Descriptions of growth on agar slants.

5. Using the other broth culture, inoculate a broth culture, agar slant, and semisolid agar deep, as described in steps 2, 3, and 4, using your inoculating *needle*.

6. To transfer *Proteus vulgaris,* flame your loop and allow it to cool. Flame the mouth of the tube and carefully scrape a small amount of the culture from the agar. Flame the mouth of the tube and replace the cap. Inoculate a broth and slant as described in

steps 2 and 3. Inoculate a semisolid agar deep with an inoculating needle as described in step 4. Label all inoculated tubes properly.

7. Incubate all tubes at 35°C until the next period.

8. Record the appearance of each culture, referring to Figure 10.6.

9. Make a smear of the *Streptococcus* broth culture and the *Streptococcus* slant culture. Gram stain both smears and compare them.

EXERCISE 10

Transfer of Bacteria: Aseptic Technique

Name _____

Date _____

Lab Section _____

Purpose _____

_____

_____

Data

Nutrient Broth

Describe the nutrient broth cultures.

Bacterium	Is it turbid?	Is flocculant, pellicle, or sediment present?	Pigment?
Streptococcus lactis			
Pseudomonas aeruginosa			
Proteus vulgaris			

Nutrient Agar Slant

Sketch the appearance of each culture. Note any pigmentation.

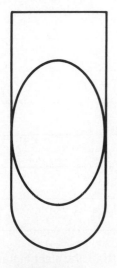

Streptococcus lactis

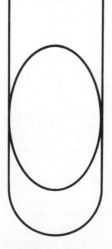

Pseudomonas aeruginosa

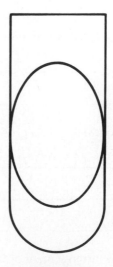

Proteus vulgaris

Type of growth:

_____   _____   _____

Nutrient Agar Deep

Show the location of bacterial growth and note any pigment formation.

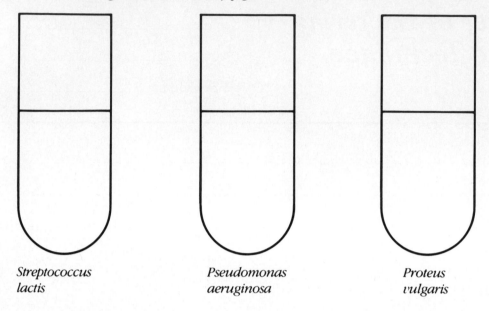

| Streptococcus | Pseudomonas | Proteus |
| lactis | aeruginosa | vulgaris |

Comparison of Broth and Slant Cultures

	Streptococcus lactis	
	Original Broth Culture	Slant Culture
Gram Stain		
Morphology		
Arrangement		

Questions

1. Did growth occur at different levels in the agar deep? _____

    _____

    _____

2. Were any of the bacteria growing in the semisolid agar deeps motile? Explain. _____

    _____

    _____

    _____

3. Why was the arrangement of *Streptococcus* from the broth culture different from the slant culture in the

    second period? _____

    _____

4. What is the primary use of slants? Of deeps? Of broths? _____

_____

_____

5. Can you determine whether a broth culture is pure (all one species) by visually inspecting it without a

microscope? _____ An agar deep culture? _____

An agar slant culture? _____

6. When is a loop preferable for transferring bacteria? Use an illustration from your results to answer. When is a

needle preferable? _____

_____

_____

7. What is the purpose of flaming the loop before use? After use? _____

_____

_____

8. Why must the loop be cool before touching it to a culture? Should you set it down to let it cool? How do you

determine when it is cool? _____

_____

_____

9. Why is aseptic technique important? _____

_____

_____

10. How can you tell that the media provided for this exercise were sterile? _____

_____

_____

_____

Isolation of Bacteria by Dilution Techniques

*Experience is the father of wisdom,
and memory the mother.*

THOMAS FULLER

Objectives

After completing this exercise you should be able to:

1. Isolate bacteria by the streak plate and pour plate techniques.
2. Prepare and maintain a pure culture.

Background

In nature most microbes are found growing in environments that contain many different organisms. Unfortunately, mixed cultures are of little use in studying microorganisms because of the difficulty they present in determining which organism is responsible for any observed activity. A **pure culture,** one containing a single kind of microbe, is required in order to study concepts such as growth characteristics, pathogenicity, metabolism, and antibiotic susceptibility. Since bacteria are too small to separate directly without sophisticated micromanipulation equipment, indirect methods of separation must be used.

In the 1870s, Joseph Lister attempted to obtain pure cultures by performing serial dilutions until each of his containers theoretically contained one bacterium. However, success was very limited, and **contamination,** the presence of unwanted microorganisms, was common. In 1880 Koch prepared solid media, after which microbiologists could separate bacteria by dilution and trap them in the solid media. An isolated bacterium grows into a visible colony that consists of one kind of bacterium.

Currently there are three dilution methods commonly used for the isolation of bacteria: the streak plate, the spread plate, and the pour plate. In the **streak plate technique,** a loop is used to streak the mixed sample many times over the surface of a solid culture medium in a Petri plate. Theoretically, by streaking the loop repeatedly over the agar surface, the bacteria fall off the loop one by one and each cell develops into a colony. The streak plate is the most common isolation technique in use today.

The spread plate and pour plate are quantitative techniques that allow determination of the number of bacteria in a sample. In the **spread plate technique,** a small amount of a previously diluted specimen is spread over the surface of a solid medium using a bent rod (shaped like a hockey stick). The spread plate technique will be used in Exercise 27.

In the **pour plate technique,** a small amount of diluted sample is mixed with melted agar and poured into Petri plates. After incubation, bacterial growth is visible as colonies in and on the agar of a pour plate. To determine the number of bacteria in the original sample, a plate with between 25 and 250 colonies is selected. Fewer than 25 colonies is inaccurate because a single contaminant causes at least a 4% error. A plate with greater than 250 colonies is difficult to count. The number of bacteria in the original sample is calculated using the following equation:

$$\text{Bacteria per ml} = \frac{\text{Number of colonies}}{\text{Dilution*}}$$

Materials

Petri plates containing nutrient agar (2)

Tubes containing melted nutrient agar (3)

Sterile Petri plates (3)

Nutrient agar slant

250-ml beaker

Sterile 1-ml pipettes (3)

Propipette or pipet bulb

---

*Dilution refers to the dilution of sample (Appendix B). For example, if 37 colonies were present on the 1:100 plate, the calculation would be

$$\text{Bacteria per ml} = \frac{37}{10^{-2}} = 37 \times 10^2 = 3700 = 3.7 \times 10^3$$

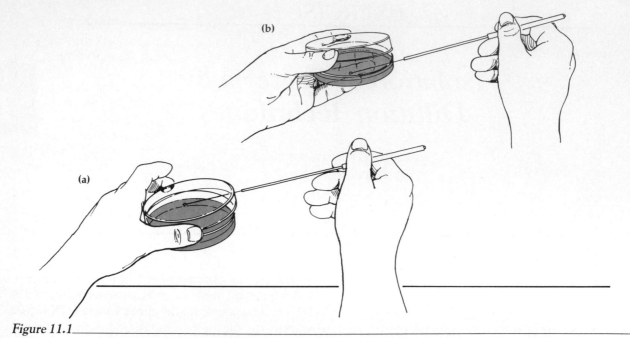

Figure 11.1_____

Inoculation of a solid medium in a Petri plate. Lift one edge of the cover while **(a)** the plate rests on the table, or **(b)** is held.

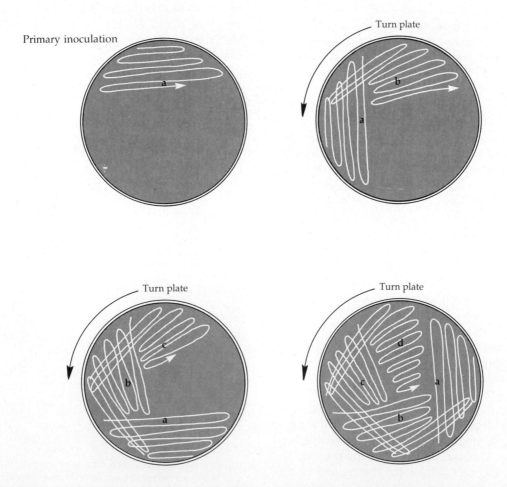

Figure 11.2 _____

Streak plate technique for pure culture isolation of bacteria. The direction of streaking is indicated by the arrows. Between each section, sterilize the loop and reinoculate with a fraction of the bacteria by going back across the previous section.

Cultures

Mixed broth culture of bacteria

Turbid nutrient broth from Exercise 9

Techniques Required

Exercises 9 and 10; Appendices A and B.

Procedure

Streak Plate

1. Label the bottoms of two nutrient agar plates to correspond to the two broth cultures: mixed culture and turbid broth from Exercise 9.
2. Flame the inoculating loop to redness, allow it to cool, and aseptically obtain a loopful of one broth culture.
3. The streaking procedure may be done with the Petri plate on the table (Figure 11.1*a*) or held in your hand (Figure 11.1*b*).

a. To streak a plate (Figure 11.2), lift one edge of the Petri plate cover and streak the first sector by making as many streaks as possible without overlapping previous streaks.

b. Flame your loop and let it cool. Turn the plate so the next sector is on top. Streak through one area of the first sector, then streak a few times away from the first sector.

c. Flame your loop, turn the plate again, and streak through one area of the second sector, and streak the third sector.

d. Flame your loop, streak through one area of the third sector, and streak the remaining area of the agar surface. Flame your loop before setting it down. Why? _____

_____

4. Streak two plates: one of the mixed culture provided and one of the turbid broth from Exercise 9. Label each plate on the bottom with your name, lab section, date, and source of the inoculum.
5. Incubate the plates in an inverted position in the

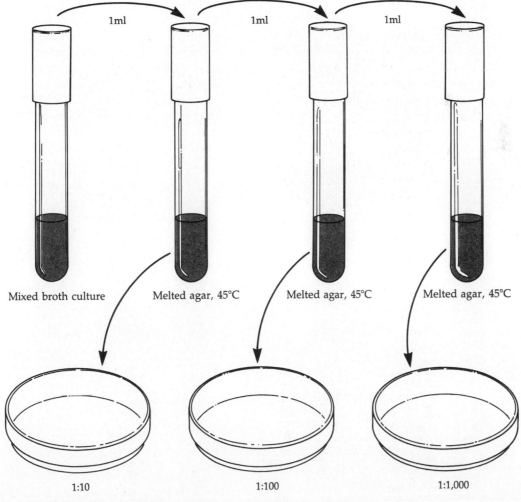

1ml   1ml   1ml

Mixed broth culture   Melted agar, 45°C   Melted agar, 45°C   Melted agar, 45°C

1:10   1:100   1:1,000

Figure 11.3 _____

Pour plate technique. Bacteria are diluted through a series of tubes containing melted nutrient agar. The agar and bacteria are poured into sterile Petri plates. The bacteria will form colonies where they were trapped in the agar.

35°C incubator until discrete, isolated colonies develop (usually 24 to 48 hours). Why inverted? ____

6. After incubation, record your results. (Refer to Figure 9.1 and color plates VI.1 and VI.2.)
7. Prepare a subculture of one colony. Sterilize your needle by flaming. Let it cool. Why use a needle instead of a loop? ____
To subculture, touch the center of a small isolated colony and then aseptically streak a sterile nutrient agar slant (Figure 10.4). How can you tell whether or not you touched only one colony and whether or not you have a pure culture? ____

8. Incubate the slant at 35°C until good growth is observed. Describe the growth pattern (Figure 10.6).

Pour Plate (Figure 11.3)
1. Label the bottoms of three sterile Petri plates with your name, lab section, and date. Label one plate 1:10, another, 1:100, and the third one, 1:1,000.
2. Fill a beaker with hot (45–50°C) water (about 3–6 cm) and place three tubes of melted nutrient agar in the beaker.
3. Select a mixed broth culture. Place the labeled plates on your workbench right-side up.
4. Aseptically transfer 1 ml of the broth to a tube of melted agar. Mix well, as shown in Figure 11.4. Using a different pipette, transfer 1 ml from this tube to a second tube. Aseptically pour the contents of the first tube into Petri plate 1:10.
5. Mix the second tube. With the third pipette, aseptic-

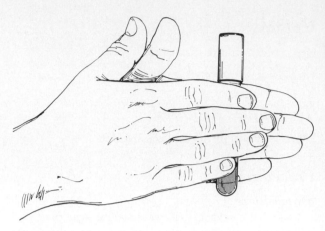

Figure 11.4 _____
Mix the inoculum in a tube of melted agar by rolling the tube between your hands.

ally transfer 1 ml to the third tube. Pour the contents of the second tube into plate 1:100. Mix the third tube and pour into the remaining plate, 1:1,000.
6. Discard the tubes. Let the agar harden in the plates, then incubate at 35°C in an inverted position until growth is seen. *Suggestion:* When incubating multiple plates, use a rubber band to hold them together.
7. After incubation, count the number of colonies on the plates. Remember that more than 250 is too many to count and less than 25 is too few to count.

EXERCISE 11

Isolation of Bacteria by Dilution Techniques

Name _____

Date _____

Lab Section _____

Purpose _____

_____

_____

Results

Streak Plate

Sketch the appearance of the streak plates.

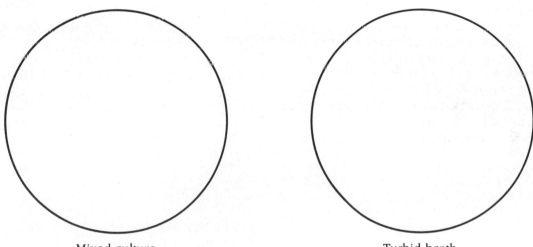

Mixed culture   Turbid broth

Fill in the following table using colonies from the most isolated streak areas.

Culture	Colony Description (Each different appearing colony should be described.)				
	Diameter	Appearance	Margin	Elevation	Color
Mixed culture					
Turbid broth					

Pour Plate

Dilution	Number of Colonies
1:10	
1:100	
1:1,000	

_____ Bacteria/ml

Calculate the number of bacteria/ml in the mixed culture. Show your calculations.

Subculture

Describe the growth on your slant. _____

Questions

1. How many different bacteria were in the mixed culture? _____ How many in the

    turbid broth? _____ How can you tell? _____

    _____

2. How do the colonies on the surface of the pour plate differ from the colonies suspended in the agar? _____

    _____

    _____

3. What is a contaminant? _____

4. How would you determine whether a colony was a contaminant on a streak plate? _____

    _____

    On a pour plate? _____

5. What would happen if the plates were incubated a week longer? _____

    A month? _____

6. Could some bacteria grow on the streak plate and not be seen using the pour plate technique? _____

    Explain. _____

    _____

7. What is a disadvantage of the streak plate technique? Of the pour plate technique? _____

    _____

    _____

Special Media for Isolating Bacteria

Objectives

After completing this exercise you should be able to:

1. Differentiate between selective and differential media.
2. Provide an application for enrichment and selective media.

Background

One of the major limitations of dilution techniques used to isolate bacteria is that organisms present in limited amounts may be diluted out on plates filled with dominant bacteria. For example, if the culture to be isolated has 1 million of bacterium A and only 1 of bacterium B, bacterium B will probably be limited to the first sector in a streak plate. To help isolate organisms found in the minority, various enrichment and selective culturing methods are available that enhance the growth of some organisms and inhibit the growth of other organisms. **Selective media** contain chemicals that prevent the growth of unwanted bacteria without inhibiting the growth of the desired organism. **Enrichment media** contain chemicals that enhance the growth of desired bacteria. Other bacteria will grow, but the growth of the desired bacteria will be increased.

Another category of media useful in identifying bacteria are **differential media.** These media contain various nutrients that allow the investigator to distinguish one bacterium from another by how they metabolize or change the media. Differential media will be used in later exercises (see, for example, Part 4).

Because multiple methods and multiple media exist, you must be able to match the correct procedure to the desired microbe. In this exercise, we will determine one criterion used to select a culture medium. For example, if bacterium B is salt-tolerant, salt could be added to the culture medium. Physical conditions can also be used to select for a bacterium. If bacterium B is heat-resistant, the specimen could be heated before isolation. Two culture media will be compared in this exercise (Table 12.1).

Materials

Petri plate containing phenylethyl alcohol agar

Petri plate containing tryptose agar

Gram staining reagents

Cultures

Staphylococcus epidermidis

Escherichia coli

Mixed culture of *Escherichia* and *Staphylococcus*

Techniques Required

Exercises 5 and 10.

Table 12.1
Chemical Composition of Media Used in This Exercise

Tryptose Agar		Phenylethyl Alcohol Agar	
Tryptose	1.0%	Tryptose	1.0%
Beef extract	0.3	Beef extract	0.3
NaCl	0.5	Phenylethyl alcohol	0.25
Agar	1.5	NaCl	0.5
		Agar	1.5

Procedure

1. Using a marker, divide each plate into three sections by labeling the bottoms. Label one section on each plate for each culture.
2. Streak each culture on the agar, as shown in Figure 12.1.
3. Incubate the plates in an inverted position at 35°C. Record the results after 24 hours, and after 48 to 72 hours, of incubation. Gram stain different appearing colonies from each sector.

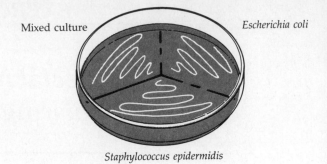

Mixed culture

Escherichia coli

Staphylococcus epidermidis

Figure 12.1

Divide a Petri plate into three sections by drawing lines on the bottom of the plate. Inoculate each section by streaking with an inoculating loop.

EXERCISE 12

Special Media for Isolating Bacteria

Name _____

Date _____

Lab Section _____

Purpose _____

_____

_____

Data

Organism	Amount of Growth				Gram Stain Results
	Tryptose Agar		Phenylethyl Alcohol Agar		
	24 hr	48–72 hr	24 hr	48–72 hr	

Questions

1. What do your results indicate? What difference did you see between 24 and 48 to 72 hours? _____

_____

_____

_____

2. Is phenylethyl alcohol agar a selective or differential medium? _____

_____

_____

3. How did the media composition affect bacterial growth? _____

_____

_____

4. Design another medium that would select for the same group of organisms. How is your medium selective? ___

_____

_____

5. Design an enrichment medium to isolate a detergent degrading bacterium that is found in soil. _____

_____

_____

_____

_____

_____

_____

PART 4

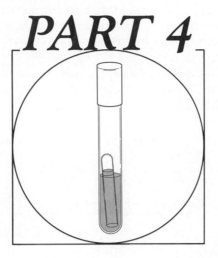

Microbial Metabolism

---------------------------------- *EXERCISES* ----------------------------------

Leeuwenhoek saw microorganisms in wine, and Pasteur demonstrated that the microbes were living organisms. In 1872 Pasteur wrote:

> *It is impossible that the organic matter of the newly formed ferments [microorganisms] contain a single carbon atom which has not been derived from the fermented substance.**

The chemical reactions observed by Pasteur and the chemical reactions that occur within all living organisms are referred to as **metabolism.** Metabolic processes involve **enzymes,** which are proteins that catalyze biologic reactions. The majority of enzymes function inside a cell; that is, they are **endoenzymes.** Many bacteria make some enzymes, called **exoenzymes,** which are released from the cell to catalyze reactions outside of the cell (see the figure).

Some bacteria use particular metabolic pathways in the presence of oxygen **(aerobic)** and other pathways in the absence of oxygen **(anaerobic).** Pasteur continued:

*Quoted in H. A. Lechevalier and M. Solotorovsky. Three Centuries of Microbiology. New York: Dover Publications, 1974, p. 26.

I have demonstrated that since this ferment [microorganism] survived in the presence of some free oxygen, it lost its fermentative abilities in proportion to the concentration of this gas.

Because many bacteria share the same colony and cell morphology, additional factors, such as metabolism, are used to characterize and classify them. On the basis of which substrates a particular bacterium uses and which metabolic products it forms, laboratory tests have been designed to determine which enzymes the bacterium has.

The first five exercises in Part 4 introduce concepts in microbial metabolism and laboratory tests used to detect various metabolic activities. In Exercise 18, unknown bacteria will be identified on the basis of metabolic characteristics.

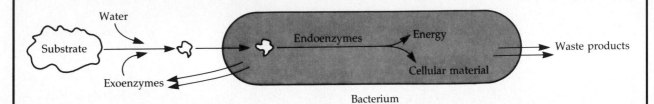

Large molecules are broken down outside of a cell by exoenzymes. Smaller molecules released by this reaction are taken into the cell and further degraded by endoenzymes.

Carbohydrate Catabolism

*The men who
try to do something and fail are infinitely
better than those who try nothing and
succeed.*

LLOYD JONES

Objectives

After completing this exercise you should be able to:

1. Define the following terms: carbohydrate, catabolism, and exoenzyme.
2. Differentiate between oxidative and fermentative catabolism.
3. Perform and interpret microbial starch hydrolysis and OF tests.

Background

Chemical reactions that release energy from the decomposition of complex organic molecules are referred to as **catabolism.** Most bacteria catabolize carbohydrates for carbon and energy. **Carbohydrates** are organic molecules that contain carbon, hydrogen, and oxygen in the ratio $(CH_2O)_n$. Carbohydrates can be divided into three groups based on size: monosaccharides, disaccharides, and polysaccharides. **Monosaccharides** are simple sugars containing from three to seven carbon atoms, **disaccharides** are composed of two monosaccharide molecules, and **polysaccharides** consist of eight or more monosaccharide molecules.

Exoenzymes are mainly **hydrolytic enzymes** that break down, by the addition of water, large substrates into smaller components that can be transported into the cell. The exoenzyme amylase hydrolyzes the polysaccharide starch into smaller carbohydrates. Glucose, a monosaccharide, can be released by hydrolysis (Figure 13.1). In the laboratory, the presence of an exoenzyme is determined by looking for a change in the substrate outside of a bacterial colony.

Glucose can enter a cell and be catabolized; some bacteria catabolize glucose oxidatively, producing carbon dioxide and water. **Oxidative catabolism** requires the presence of molecular oxygen (O_2). Most bacteria, however, ferment glucose without using oxygen. **Fermentative catabolism** does not require oxygen but may occur in the presence of oxygen. The

metabolic end-products of fermentation are small organic molecules, usually organic acids. Some bacteria are both oxidative and fermentative; others are neither but obtain their carbon and energy by other means.

Whether an organism is oxidative or fermentative can be determined by using Rudolph Hugh and Einar Leifson's OF basal media with the desired carbohydrate added. **OF medium** is a nutrient semisolid agar deep containing a *high* concentration of carbohydrate and a *low* concentration of peptone. The peptone will support the growth of nonoxidative–nonfermentative bacteria. Two tubes are used: one open to the air and one sealed to keep air out. OF medium contains the indicator bromthymol blue, which turns yellow in the presence of acids, indicating catabolism of the carbohydrate. Alkaline conditions, due to the use of peptone and not the carbohydrate, are indicated by a dark blue color. If the carbohydrate is metabolized in both tubes, fermentation has occurred. Some bacteria produce gases from the fermentation of carbohydrates. Gases can be seen as pockets or cracks in the OF medium. An organism that can only use the carbohydrate oxidatively will produce acid in the open tube only.

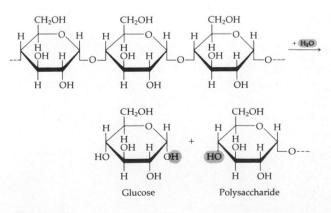

Figure 13.1 _____

Starch hydrolysis. A molecule of water is used when starch is hydrolyzed.

Carbohydrate catabolism will be demonstrated in this exercise. These differential tests will be very important in identifying bacteria in later exercises.

Materials

Petri plate containing nutrient starch agar

OF-glucose deeps (2)

Mineral oil

Gram's iodine

Cultures (as assigned)

Bacillus subtilis

Escherichia coli

Pseudomonas aeruginosa

Alcaligenes faecalis

Techniques Required

Exercise 10.

Procedure

Starch Hydrolysis

1. With a marker, divide the starch agar into three sectors by labeling the bottom of the plate.
2. Streak a single line of *Bacillus, Escherichia,* and *Pseudomonas.*
3. Incubate, inverted, at 35°C until the next lab period.
4. Record any bacterial growth, then flood the plate with Gram's iodine (Figure 13.2). Areas of starch hydrolysis will appear clear, while unchanged starch will stain dark blue. Record your results.

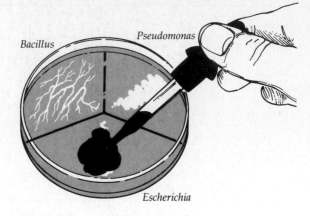

Figure 13.2 _____

Starch hydrolysis test. After incubation, add iodine to the plate to detect the presence of starch.

OF-Glucose

1. Using an inoculating needle, inoculate two tubes of OF-glucose media with the assigned bacterial culture *(Escherichia, Pseudomonas,* or *Alcaligenes).*
2. Pour about 5 mm of mineral oil over the medium in one of the tubes.
3. Incubate both tubes at 35°C until the next lab period.
4. Observe the tubes and record the following: the presence of growth, whether glucose was utilized, and the type of metabolism. Motility can also be ascertained from the OF tubes. How? _____
_____
5. Observe and record the results from the microorganisms you did not culture. (See color plate III.1.)

EXERCISE 13

Carbohydrate Catabolism

Name _____

Date _____

Lab Section _____

Purpose _____

_____

_____

Data

Record your results in the following tables.

Starch Hydrolysis

Organism	Growth	Color of Medium Around Colonies After Addition of Iodine	Starch Hydrolysis: (+) = yes ( ) = no
Bacillus subtilis			
Pseudomonas aeruginosa			
Escherichia coli			

OF-Glucose

Organism	Growth		Color		Gas Produced	Fermenter (F), Oxidizer (O) Neither (−)	Motile
	Open Tube	Plugged Tube	Open Tube	Plugged Tube			
Pseudomonas aeruginosa							
Alcaligenes faecalis							
Escherichia coli							

Questions

1. Which organism(s) gave a positive test for starch hydrolysis? _____

_____

How can you tell? _____

2. What would be found in the clear area that would not be found in the blue area of a starch agar plate after the

addition of iodine? _____

_____

3. How can you tell amylase is an exoenzyme and not an endoenzyme? _____

_____

4. How can you tell from OF-glucose medium whether an organism oxidizes glucose? _____

_____

Ferments glucose? _____

Doesn't use glucose? _____

5. If an organism grows in the OF-glucose medium that is exposed to air, is the organism oxidizing or fermenting

glucose? Explain. _____

_____

6. How can organisms that don't utilize starch grow on a starch agar plate? _____

_____

_____

7. If iodine were not available, how would you determine if starch hydrolysis had occurred? _____

_____

8. Locate the H^+ and OH^- ions from the water molecule that was split (hydrolyzed) in this reaction, showing
   esculin hydrolysis.

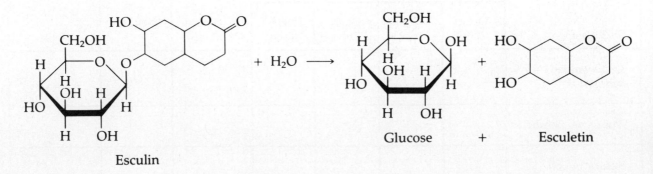

Fermentation of Carbohydrates

Objectives

After completing this exercise you should be able to:

1. Define fermentation.
2. Perform and interpret carbohydrate fermentation tests.
3. Perform and interpret the MR and V–P tests.

Background

Once a bacterium has been determined to be fermentative by the OF test (Exercise 13), further tests can determine which carbohydrates are fermented; in some instances, the end-products can also be determined. Many carbohydrates, including monosaccharides such as glucose, disaccharides like sucrose, and polysaccharides such as cellulose, can be fermented. Many bacteria produce organic acids (for example, lactic acid) and hydrogen and carbon dioxide gases from carbohydrate fermentation. A **fermentation tube** is used to detect acid and gas production from carbohy-

drates. The fermentation medium contains peptone, an acid–base indicator, an inverted tube to trap gas, and 0.5% to 1.0% of the desired carbohydrate. In Figure 14.1, phenol red indicator is red (neutral) in an uninoculated fermentation tube; fermentation that results in acid production will turn the indicator yellow (pH of 6.8 or below). When gas is produced during fermentation, some will be trapped in the inverted tube. Fermentation occurs with or without oxygen present; however, during prolonged incubation periods (greater than 48 hours), many bacteria will begin growing oxidatively on the peptone after exhausting the carbohydrate supplied, causing neutralization of the indicator.

Fermentation processes can produce a variety of end-products, depending on the substrate, the incubation, and the organism. In some instances, large amounts of acid may be produced, while in others, a majority of neutral products may result (Figure 14.2a). The **MRVP test** is used to distinguish between organisms that produce large amounts of acid from glucose and those that produce the neutral product *acetoin*. MRVP medium is a glucose-supplemented nutrient broth used for the **methyl red (MR) test** and the **Voges–Proskauer (V–P) test.** If an organism produces a large amount of organic acid from glucose, the medium will remain red when methyl red is added, indicating the pH is below 4.4. If neutral products are produced, methyl red will turn yellow, indicating a pH above 6.0. The production of acetoin is detected by the addition of potassium hydroxide and α-naphthol. If acetoin is present, the upper part of the medium will turn red; a negative V–P test will turn the medium light brown. The chemical process is shown in Figure 14.2b. The production of acetoin is dependent on the length of incubation and the environmental conditions, as shown in Figure 14.3.

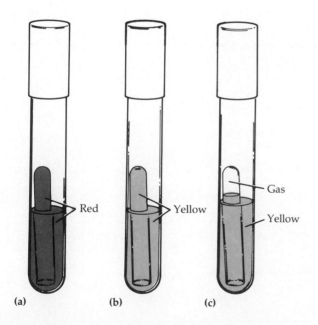

(a)   **(b)**   **(c)**

Figure 14.1 _____
Carbohydrate fermentation tube. **(a)** The phenol red indicator is red in a neutral or alkaline solution. **(b)** Phenol red turns yellow in the presence of acids. **(c)** Gases are trapped in the inverted tube, while the indicator shows the production of acid.

Materials

Glucose fermentation tube

Lactose fermentation tube

Sucrose fermentation tube

MRVP broths (4)

Methyl red

V–P reagent I, α-naphthol solution

V–P reagent II, potassium hydroxide (40%)

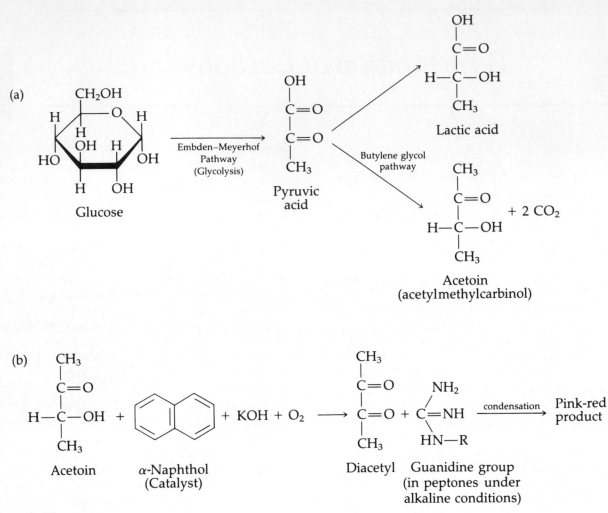

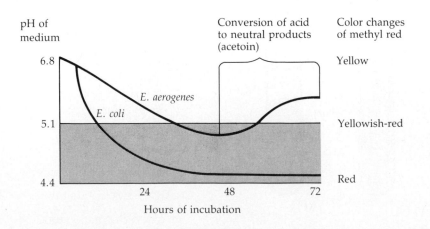

Figure 14.2

Fermentation of glucose. **(a)** Organic acids, such as lactic acid, or neutral products, such as acetoin, may result from fermentation. **(b)** Potassium hydroxide (KOH) and α-naphthol are used to detect acetoin.

Figure 14.3

The production of acetoin is dependent on incubation time and pH.

Cultures (as assigned)

Escherichia coli

Enterobacter aerogenes

Alcaligenes faecalis

Proteus vulgaris

Techniques Required

Exercises 10 and 13.

Procedure

Fermentation Tubes

1. Use a loop to inoculate the fermentation tubes with the assigned bacterial culture.
2. Incubate at 35°C. Examine at 24 and 48 hours for growth, acid, and gas. Why is it important to record the presence of growth? _____
   _____
3. Record your results with this culture, as well as your results for the other species tested. (See color plate III.2.)

MRVP Tests

1. Using a loop, inoculate two MRVP tubes with *Escherichia* and two with *Enterobacter*.
2. Incubate at 35°C for 48 hours or longer. Why is time of incubation important? _____
   _____
3. To one tube of each set *(Escherichia* and *Enterobacter)*, add 5 drops of methyl red. Record the resulting color. Red is a positive methyl red test.
4. To the other set of two tubes add 0.6 ml (12 drops) of V–P reagent I and 0.2 ml (4 drops) of V–P reagent II.
5. Gently shake the tubes with the caps off to expose the media to oxygen in order to oxidize the acetoin.
6. Allow the tubes to stand for 15–30 minutes. A positive V–P test will develop pink to red color.
7. Record your results. (See color plates III.6 and III.7.)

EXERCISE 14

Fermentation of Carbohydrates

Name _____

Date _____

Lab Section _____

Purpose _____

_____

_____

Data

Fermentation Tubes

Organism		Carbohydrate											
		Glucose				Lactose				Sucrose			
		Growth	Color	Acid	Gas	Growth	Color	Acid	Gas	Growth	Color	Acid	Gas
Escherichia coli	24 hrs.												
	48 hrs.												
Enterobacter aerogenes	24 hrs.												
	48 hrs.												
Alcaligenes faecalis	24 hrs.												
	48 hrs.												
Proteus vulgaris	24 hrs.												
	48 hrs.												

MRVP Tests

Organism	Growth	MR		V–P	
		Color	+ or −	Color	+ or −
Escherichia coli					
Enterobacter aerogenes					

Questions

1. Why are fermentation tubes evaluated at 24 and 48 hours? _____

   _____

   What would happen if an organism used up all the carbohydrate in a fermentation tube? _____

   _____ What would it

   use for energy? _____ What color would the indicator be then? _____

2. Could an organism be MR and V–P positive? Explain. _____

   _____

   _____

3. If an organism oxidatively metabolizes glucose, what result will occur in the fermentation tubes? _____

   _____

   _____

4. Were these media differential or selective? _____

5. How could you determine whether a bacterium fermented the following carbohydrates: mannitol, sorbitol,

   adonitol, or arabinose? _____

   _____

   _____

6. If a bacterium cannot ferment glucose, why not test its ability to ferment other carbohydrates? _____

   _____

   _____

   _____

Protein Catabolism, Part 1

Objectives

After completing this exercise you should be able to:

1. Determine a bacterium's ability to hydrolyze gelatin.
2. Test for the presence of urease.
3. Perform and interpret a litmus milk test.

Background

Proteins are organic molecules that contain carbon, hydrogen, oxygen, and nitrogen; some proteins also contain sulfur. The subunits that make up a protein are called **amino acids** (Figure 15.1). Amino acids bond together by **peptide bonds** (Figure 15.2), forming a small chain (a **peptide**) or a larger molecule (a **polypeptide**).

Bacteria can use a variety of proteins and amino acids as carbon and energy sources when carbohydrates are not available. However, amino acids are primarily used in anabolic reactions.

Large protein molecules, such as gelatin, are hydrolyzed by exoenzymes, and the smaller products of hy-

drolysis are transported into the cell. Hydrolysis of gelatin can be demonstrated by growing bacteria in nutrient gelatin. Nutrient gelatin dissolves in warm water (50°C), solidifies **(gels)** when cooled below 25°C, and liquefies **(sols)** when heated to about 25°C. When an exoenzyme hydrolyzes gelatin, it liquefies and does not solidify even when cooled below 20°C.

Some bacteria can hydrolyze the protein in milk called **casein.** Casein hydrolysis can be detected in litmus milk. Litmus milk consists of skim milk and the indicator litmus. The medium is opaque, due to casein in colloidal suspension, and the litmus is blue. After **peptonization** (hydrolysis of the milk proteins), the medium becomes clear as a result of hydrolysis of casein to soluble amino acids and peptide fragments. Litmus milk is also used to detect **lactose fermentation,** since litmus turns pink in the presence of acid. Excessive amounts of acid will cause **coagulation** (curd formation) of the milk. **Catabolism of amino acids** will result in an alkaline (purple) reaction. In addition, some bacteria can **reduce litmus** (Exercise 17), causing the litmus indicator to turn white in the bottom of the tube.

Urea is a waste product of protein digestion in most vertebrates and is excreted in the urine. Presence of the enzyme **urease,** which liberates ammonia from urea (Figure 15.3), is a useful diagnostic test for identifying bacteria. **Urea agar** contains peptone, glucose, urea, and phenol red. The pH of the prepared medium is 6.8 (phenol red turns yellow). During incubation, bacteria possessing urease will produce ammonia that raises the pH of the medium, turning the indicator fuchsia (hot pink) at pH 8.4.

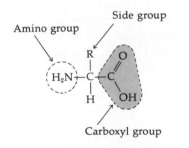

Figure 15.1 _____

General structural formula for an amino acid. The letter R stands for any of a number of groups of atoms. Different amino acids have different R groups.

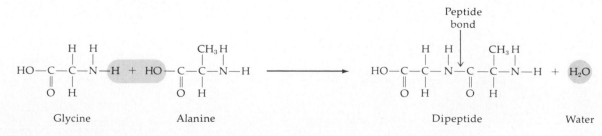

Figure 15.2 _____

Peptide bond formation. The amino acids glycine and alanine combine to form a dipeptide. The newly formed bond between the nitrogen atom of glycine and the carbon atom of alanine is called a peptide bond.

We will investigate bacterial action on nutrient gelatin, litmus milk, and urea agar in this exercise, and amino acid metabolism in Exercise 16.

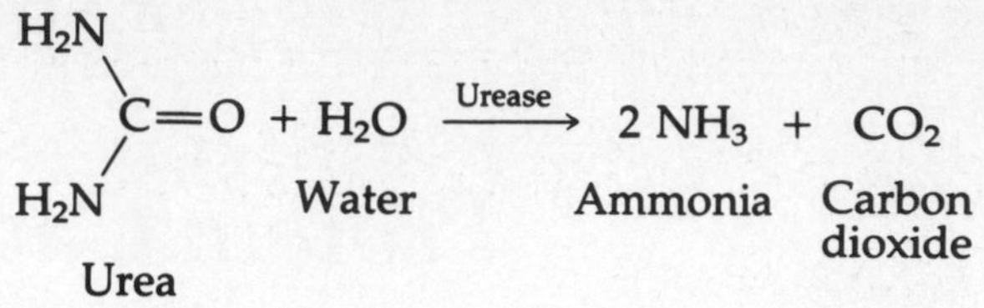

Materials

Tubes containing nutrient gelatin (2)

Tubes containing litmus milk (2)

Tubes containing urea agar (2)

Cultures

Pseudomonas aeruginosa

Proteus vulgaris

Techniques Required

Exercise 10.

Procedure

1. Label one tube of each medium *"Pseudomonas"* and the other tube *"Proteus."*
2. Gelatin hydrolysis. Examine the nutrient gelatin: Is it solid or liquid? _____
What is the temperature of the laboratory? _____
If the gelatin is solid, what would you need to do to liquefy it? _____
To resolidify it? _____
   a. Inoculate one tube with *Pseudomonas* and one with *Proteus*.
   b. Incubate at room temperature and record your observations at 2 to 4 days and at 4 to 7 days. Do not agitate the tube when the gelatin is liquid. Why? _____

$$H_2N \diagdown \atop H_2N \diagup C{=}O + H_2O \xrightarrow{\text{Urease}} 2\,NH_3 + CO_2$$

Urea   Water   Ammonia   Carbon dioxide

Figure 15.3 _____
Urea hydrolysis.

   c. If the gelatin has liquefied, place the tube in a beaker of crushed ice for a few minutes. Is the gelatin still liquefied? _____
Record your results. Indicate liquefaction or hydrolysis by (+). (See color plate III.4.)
3. Litmus milk. Describe the appearance of litmus milk. _____
_____
   a. Inoculate one tube with *Pseudomonas* and one with *Proteus*.
   b. Incubate the litmus milk at 35°C for 24 to 48 hours and record the results. Litmus is pink under acidic conditions, purple under alkaline conditions, and white when reduced. (See color plate III.3.)
4. Urease test.
   a. Inoculate one urea agar slant with *Pseudomonas* and one with *Proteus*.
   b. Incubate for 24 to 48 hours at 35°C. (See color plate III.11.) Record your results: (+) for the presence of urease (red) and (−) for no urease. What color is phenol red at pH 6.8 or below? _____
At pH 8.4 or above? _____

EXERCISE 15

Protein Catabolism, Part 1

Name _____

Date _____

Lab Section _____

Purpose _____

_____

_____

Data

Fill in the following table.

Test	Results	
	Pseudomonas aeruginosa	*Proteus vulgaris*
Gelatin hydrolysis at _____ days growth (+ or −)		
at _____ days growth (+ or −)		
Litmus milk Peptonization		
Acid		
Alkaline		
Coagulation		
Reduction		
No change		
Urea agar growth (+ or −)		

Diagram the appearance of the gelatin.

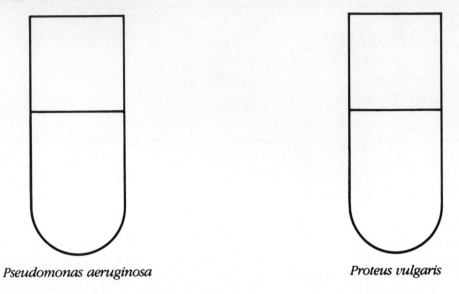

Pseudomonas aeruginosa   *Proteus vulgaris*

Questions

1. Nutrient gelatin can be incubated at 35°C. What would have to be done to determine hydrolysis after incubation

    at 35°C? _____

    _____

    _____

2. Why is agar used as a solidifying agent instead of gelatin? _____

    _____

    _____

3. What would you find in the liquid of hydrolyzed gelatin? _____

    _____

    _____

4. What is a disadvantage of litmus milk medium in diagnostic tests? _____

    _____

    _____

5. What happens to milk when the suspended (colloidal) proteins are hydrolyzed? _____

    _____

    _____

Protein Catabolism, Part 2

Objectives

After completing this exercise you should be able to:

1. Define the following terms: deamination and decarboxylation.
2. Explain the derivation of H_2S in decomposition.
3. Perform and interpret an indole test.

Background

Once amino acids are taken into a bacterial cell, various metabolic processes can occur. Before an amino acid can be used as a carbon and energy source, the amino group must be removed. The removal of an amino group is called **deamination.** The amino group is converted to ammonia that can be excreted from the cell. Deamination results in the formation of an organic acid. Deamination of the amino acid phenylalanine can be detected by forming a colored ferric ion complex with the resulting acid (Figure 16.1). Deamination can also be ascertained by testing for the presence of ammonia using **Nessler's reagent** (Exercise 55) which turns deep yellow in the presence of ammonia.

Various amino acids may be decarboxylated. **Decarboxylation** is the removal of carbon dioxide from an amino acid. The presence of a specific decarboxylase enzyme results in the breakdown of the amino acid with the formation of the corresponding amine, libera-

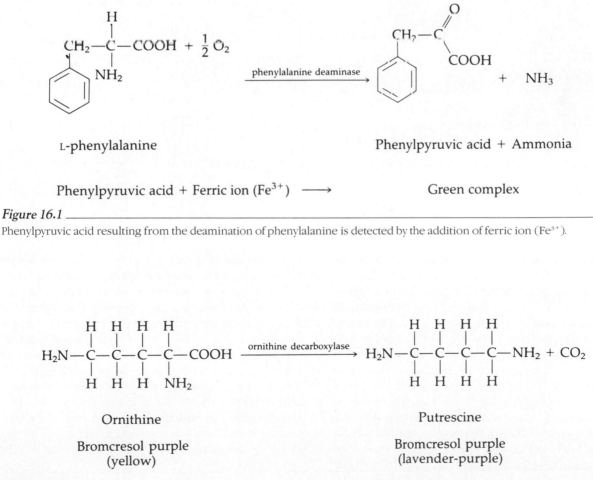

L-phenylalanine

Phenylpyruvic acid + Ammonia

Phenylpyruvic acid + Ferric ion (Fe^{3+}) $\longrightarrow$ Green complex

Figure 16.1 _____

Phenylpyruvic acid resulting from the deamination of phenylalanine is detected by the addition of ferric ion (Fe^{3+}).

Ornithine

Bromcresol purple
(yellow)

Putrescine

Bromcresol purple
(lavender-purple)

Figure 16.2 _____

Decarboxylation of the amino acid ornithine causes a change in the bromcresol purple indicator.

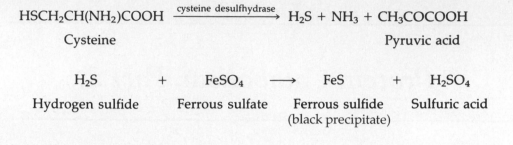

HSCH₂CH(NH₂)COOH $\xrightarrow{\text{cysteine desulfhydrase}}$ H₂S + NH₃ + CH₃COCOOH

Cysteine   Pyruvic acid

H₂S   +   FeSO₄   ⟶   FeS   +   H₂SO₄

Hydrogen sulfide   Ferrous sulfate   Ferrous sulfide   Sulfuric acid
                                                          (black precipitate)

Figure 16.3 _____

The release of H₂S from the amino acid cysteine is detected by the formation of ferrous sulfide.

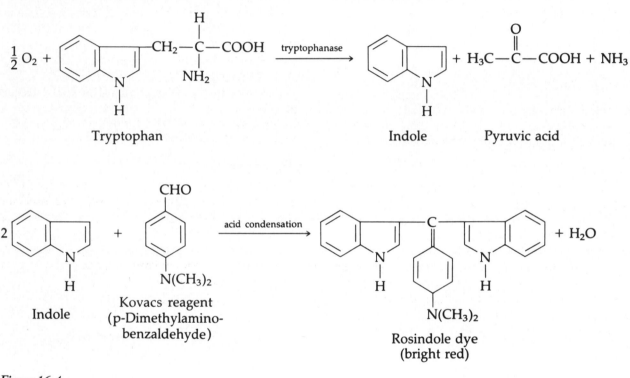

Tryptophan   Indole   Pyruvic acid

Figure 16.4 _____

Many bacteria produce indole from the amino acid tryptophan.

tion of carbon dioxide, and a shift in pH to alkaline. Media for decarboxylase reactions consist of glucose, nutrient broth, a pH indicator, and the desired amino acid. In Figure 16.2, bromcresol purple is used as a pH indicator, and a positive decarboxylase test yielding excess amines is indicated by purple. (Bromcresol purple is yellow in acidic conditions.) The names given to some of the amines, such as putrescine, indicate how foul smelling they are. Cadaverine was the name given to the foul-smelling diamine derived from decarboxylation of lysine in decomposing bodies on a battlefield.

Some bacteria liberate **hydrogen sulfide (H₂S)** from the sulfur-containing amino acids: cystine, cysteine, and methionine. H₂S can also be produced from the reduction of inorganic compounds (Exercise 17),

such as thiosulfate ($S_2O_3^{2-}$). H₂S is commonly called rotten egg gas because of the copious amounts liberated when eggs decompose. To detect H₂S production, a heavy metal salt containing ferrous ion (Fe^{2+}) is added to a nutrient culture medium. When H₂S is produced, the sulfide (S^{2-}) reacts with the metal salt to produce a visible black precipitate. The production of hydrogen sulfide from cysteine is shown in Figure 16.3.

The ability of some bacteria to convert the amino acid tryptophan to indole is a useful diagnostic tool (Figure 16.4). The **indole test** is performed by inoculating a bacterium into tryptone broth and detecting indole by the addition of dimethylaminobenzaldehyde *(Kovacs reagent)*.

Materials

Tubes containing ornithine broth (2)

Phenylalanine slants (2)

Tubes containing peptone iron deeps (2)

Tubes containing 1% tryptone broth (2)

Mineral oil

Ferric chloride reagent

Kovacs reagent

Cultures (as needed)

Escherichia coli

Pseudomonas aeruginosa

Proteus vulgaris

Enterobacter aerogenes

Techniques Required

Exercise 10.

Procedure

Ornithine Decarboxylation

1. Note the color of the ornithine broth.
2. Inoculate one tube with *Enterobacter aerogenes* and one with *Proteus vulgaris*. Add approximately 5 mm of mineral oil to the surface of each tube.
3. Incubate at 35°C for 1 to 2 days.
4. Observe for the presence of growth. (See color plate III.12.) A positive test is purple; a negative test is yellow. Why? _____

_____

Phenylalanine Deamination

1. Streak one phenylalanine slant heavily with *Proteus vulgaris* and the other with one of the remaining bacterial cultures.
2. Incubate for 1 to 2 days at 35°C. Observe for the presence of growth.
3. Add 4 or 5 drops of ferric chloride reagent to the top of the slant, allowing the reagent to run through the growth on the slant. A positive test gives a dark green color. (See color plate III.13.)

Hydrogen Sulfide Production

1. Stab one peptone iron deep with *Escherichia coli* and the other with *Proteus vulgaris*.
2. Incubate at 35°C for up to 7 days. Observe initially at 24 or 48 hours.
3. Observe for the presence of growth. Blackening in the butt of the tube indicates a positive test.

Indole Production

1. Inoculate one tryptone broth tube with *Escherichia coli* and the other with *Enterobacter aerogenes*.
2. Incubate at 35°C for 24 to 48 hours. Add 5 to 10 drops of Kovacs reagent. Agitate gently. A cherry red color in the top layer of the tube indicates a positive test.

Results

Record your results for each test.

EXERCISE 16

Protein Catabolism, Part 2

Name _____

Date _____

Lab Section _____

Purpose _____

_____

_____

Data

Fill in the following table.

	Test											
	Ornithine Decarboxylase			Phenylalanine Deaminase			Hydrogen Sulfide			Indole		
	Original color: _____			Original color: _____			Original color: _____			Original color: _____		
Organism	Growth	Color	Reaction (+ or −)	Growth	Color	Reaction (+ or −)	Growth	Color	Reaction (+ or −)	Growth	Color	Reaction (+ or −)
Escherichia coli	Not tested											
Pseudomonas aeruginosa	Not tested						Not tested			Not tested		
Proteus vulgaris										Not tested		
Enterobacter aerogenes							Not tested					

Questions

1. Why look for black precipitate (FeS) in the butt instead of on the surface of an H_2S test? _____

_____

_____

2. When spoilage of canned foods occurs, what causes the blackening of the cans? _____

_____

_____

3. Show how lysine and arginine could be decarboxylated to give the end-products indicated.

$$
\underset{\text{Lysine}}{H_2N-\overset{\displaystyle H}{\underset{\displaystyle H}{C}}-\overset{\displaystyle H}{\underset{\displaystyle H}{C}}-\overset{\displaystyle H}{\underset{\displaystyle H}{C}}-\overset{\displaystyle H}{\underset{\displaystyle H}{C}}-\overset{\displaystyle H}{\underset{\displaystyle NH_2}{C}}-COOH} \rightarrow
$$

Lysine   Cadaverine   +

_____

$$
\underset{\text{Arginine}}{\overset{\displaystyle HN}{\underset{\displaystyle NH_2}{C}}-N-\overset{\displaystyle H}{\underset{\displaystyle H}{C}}-\overset{\displaystyle H}{\underset{\displaystyle H}{C}}-\overset{\displaystyle H}{\underset{\displaystyle H}{C}}-\overset{\displaystyle H}{\underset{\displaystyle NH_2}{C}}-COOH} \rightarrow
$$

Arginine   Agmatine   +

_____

Respiration

Objectives

After completing this exercise you should be able to:

1. Define reduction.
2. Compare and contrast the following terms: aerobic respiration, anaerobic respiration, and fermentation.
3. Perform nitrate reduction, catalase, and oxidase tests.

Background

Molecules that combine with electrons liberated during metabolic processes are called **electron acceptors.** Electron acceptors become **reduced** when they gain electrons. Electrons are formed from the ionization of a hydrogen atom, as shown here:

$$H \longrightarrow H^+ + e^-$$

Hydrogen atom   Hydrogen ion   Electron

When an electron acceptor picks up an electron, it becomes negatively charged and combines with the positively charged hydrogen ion. **Reduction,** then, is a gain of electrons or hydrogen atoms.

Organic molecules act as electron acceptors in fer-

mentative metabolism. Inorganic molecules serve as electron acceptors in oxidative metabolism or **respiration.** Molecular oxygen (O_2) is the final electron acceptor in **aerobic respiration.**

In aerobic bacteria, cytochromes carry electrons to O_2. Four general classes of bacterial cytochromes have been identified, and the **oxidase test** is used to determine the presence of one of these, cytochrome c. The oxidase test is useful in identifying bacteria.

During aerobic respiration, hydrogen atoms may be combined with oxygen, forming hydrogen peroxide (H_2O_2), which is lethal to the cell. Most aerobic organisms produce the enzyme **catalase,** which breaks down hydrogen peroxide to water and oxygen, as shown here:

$$2H_2O_2 \xrightarrow{\text{catalase}} 2H_2O + O_2 \uparrow$$

Hydrogen peroxide   Water   Oxygen

In the process of **anaerobic respiration**, inorganic compounds other than O_2 act as final electron acceptors. (A few bacteria use organic electron acceptors in anaerobic respiration.) During anaerobic respiration, some bacteria reduce nitrates to nitrites; others further reduce nitrates to nitrous oxide or nitrogen gas. Other bacteria reduce nitrates to nitrites and then to ammonia.

Aerobic respiration

$$\tfrac{1}{2} O_2 + 2 H^+ + 2 e^- \longrightarrow 2 H_2O$$

Oxygen   Water

Anaerobic respiration

$$SO_4^{2-} + 10 H^+ + 10 e^- \longrightarrow H_2S + 4 H_2O$$

Sulfate ion   Hydrogen
                                     sulfide

$$CO_3^{2-} + 10 H^+ + 10 e^- \longrightarrow CH_4 + 3 H_2O$$

Carbonate ion   Methane

$$NO_3^- + 2 H^+ + 2 e^- \longrightarrow NO_2^- + H_2O \xrightarrow{6 H^+ + 6 e^-} N_2O \xrightarrow{2 H^+ + 2 e^-} N_2$$

Nitrate ion   Nitrite ion   Nitrous oxide   Nitrogen gas

Figure 17.1 _____
Chemical equations showing the reduction of some electron acceptors in bacterial respiration.

Nitrate broth (nutrient broth plus 0.1% potassium nitrate) is used to determine a bacterium's ability to reduce nitrates. Nitrites are detected by the addition of dimethyl-α-naphthylamine and sulfanilic acid to nitrate broth. A negative test (no nitrites) is further checked for the presence of nitrate in the broth by the addition of zinc. If nitrates are present, reduction has not taken place. Zinc will reduce nitrate to nitrite, and a red color will appear. If nitrates are present, reduction has not taken place. If neither nitrate nor nitrite is present, the nitrogen has been reduced to nitrous oxide (N_2O) or nitrogen gas (N_2).

The reduction of some inorganic compounds in respiration is shown in Figure 17.1. In this exercise, we will examine the oxidase test, the catalase test, and the reduction of inorganic, nitrogen-containing compounds.

Materials

Tubes containing nitrate broth (3)

Petri plates containing trypticase soy agar (2)

Nitrate reagent A (dimethyl-α-naphthylamine)

Nitrate reagent B (sulfanilic acid)

Hydrogen peroxide, 3%

Oxidase reagent or oxidase disk

Zinc dust

Cultures

Bacillus subtilis

Bacillus megaterium

Pseudomonas aeruginosa

Streptococcus lactis

Techniques Required

Exercises 10, 11, and 13.

Procedure

Nitrate Reduction Test

1. Label three tubes of nitrate broth and inoculate one with *Streptococcus lactis*, one with *Pseudomonas aeruginosa*, and one with *B. subtilis*.
2. Incubate at 35°C for 2 days.
3. Add 5 drops of nitrate A and 5 drops of nitrate B to each tube. Shake gently.
4. A red color within 30 seconds is a positive test. If red, what compound is present? _____
   _____
5. If it does not turn red, add a small pinch of zinc dust; if it turns red now, the test is negative. If not, it is positive for nitrate reduction. Why? _____
   _____
6. Record your results.

Oxidase Test

1. Divide one plate in half. Label one-half *"B. megaterium"* and the other *"P. aeruginosa."* Streak each organism on the appropriate half.
2. Incubate the plate inverted at 35°C for 24 to 48 hours.
3. To test for cytochrome oxidase, do one of the following:
   a. Drop oxidase reagent on the colonies and observe for a color change to pink within 1 minute, then blue to black. Oxidase-negative colonies will not change color.
   b. Moisten an oxidase disk and place it *on* colonies. Incubate the plate for 15 minutes at 35°C. The colonies of oxidase-positive bacteria will turn black. (See color plate III.9.)

Catalase Test

1. Divide the other plate in half and inoculate it with *S. lactis* and *B. subtilis*.
2. Incubate the plate inverted at 35°C for 24 to 48 hours.
3. Drop hydrogen peroxide on the colonies and observe for bubbles (catalase-positive). (See color plate X.3.) What gas is in the bubbles? _____

EXERCISE 17

Respiration

Name _____

Date _____

Lab Section _____

Purpose _____

_____

_____

Data

Nitrate Test

Organism	Color After Nitrate Reagents A and B	Color After Zinc	Gas	NO_3 Reduction?
Streptococcus lactis				
Pseudomonas aeruginosa				
Bacillus subtilis				

Oxidase Test

Organism	Color After Reagent	Oxidase Reaction
Bacillus megaterium		
Pseudomonas aeruginosa		

Catalase Test

Organism	Appearance after H_2O_2	Catalase Reaction
Streptococcus lactis		
Bacillus subtilis		

Questions

1. Define reduction. _____

_____

2. Why does hydrogen peroxide bubble when it is poured on a skin cut? _____

_____

_____

3. Differentiate between aerobic respiration and anaerobic respiration. _____

_____

_____

4. Differentiate between fermentation and anaerobic respiration. _____

_____

_____

5. Would nitrate reduction occur more often in the presence or absence of molecular oxygen? Explain. _____

_____

_____

Rapid Identification Methods

Objectives

After completing this exercise you should be able to:

1. Evaluate three methods of identifying enterics.
2. Name three advantages of the "systems approach" over conventional tube methods.

Background

The clinical microbiology laboratory must identify bacteria quickly and accurately. Accuracy is improved by using standardized tests. The IMViC tests were developed as a means of separating enterics, particularly the coliforms,* using a standard combination of four tests. Each capital letter in **IMViC** represents a test; the *i* is added for easier pronunciation. The tests are:

I for indole production from tryptophan (Exercise 16)

M for methyl red test for acid production from glucose (Exercise 14)

V for the Voges–Proskauer test for production of acetoin from glucose (Exercise 14)

C for the utilization of citrate as the sole carbon source. The Simmons citrate agar used in this exercise contains the indicator bromthymol blue. Citric acid will be the only source of carbon; therefore, only organisms capable of utilizing citric acid as a source of carbon will grow. When the citric acid is metabolized, an excess of sodium and ammonium ions results, and the indicator turns from green to blue, indicating alkaline conditions. (See color plate III.8.)

Although variation among strains does exist, IMViC reactions for selected species of enterics are given in Table 18.1.

In the last few years, **rapid identification methods** have been developed. Rapid identification systems provide a large number of results from one inoculation. Examples are Enterotube®** (see color plate III.5) and API 20E®*** for identifying oxidase-negative, gram-negative bacteria belonging to the family Enterobacteriaceae. Enterotube® is divided into twelve compartments, each containing a different substrate in agar. API 20E® consists of twenty microtubes containing dehydrated substrates. The substrates are rehydrated by adding a bacterial suspension. No culturing beyond the initial isolation is necessary with these systems. Comparisons between these rapid identification methods and conventional culture methods show that they are as accurate as conventional test-tube methods.

Computerized analysis of test results increases accuracy because each test is given a point value. Tests that are more important than others get more points. The IMViC tests are four tests of equal value.

Table 18.1
IMViC Reactions for Selected Species of Enterics

Species	Indole	Methyl red	Voges–Proskauer	Citrate
Escherichia coli	+(v)	+	−	−
Citrobacter freundii	−	+	−	+
Enterobacter aerogenes	−	−	+	+
Enterobacter cloacae	−	−	+	+
Serratia marcescens	−	+ or −*	+	+
Proteus vulgaris	+	−	−	−(v)
Proteus mirabilis	−	+	− or +**	+(v)

v = variable
* majority of strains give + results
** majority of strains give − results

*Enterics (Exercise 51) are aerobic or facultatively anaerobic, gram-negative, nonendospore-forming, rod-shaped bacteria. Coliforms are enterics that ferment lactose with acid and gas formation within 48 hours at 35°C.

**Roche Diagnostics, Division of Hoffman-LaRoche Inc., Nutley, NJ 07110.

***API Analytab Products, Division of Sherwood Medical, Plainview, NY 11803.

As commercial identification systems are developed, they can provide greater standardization in identification because they overcome the limitations of hunting through a key (Exercise 32), differences in media preparation, and evaluation of tests within a laboratory or between different laboratories. They are also time-, cost-, and labor-saving.

Materials

Petri plate containing nutrient agar

IMViC tests and reagents

API 20E® tray and reagents

Enterotube® and reagents

Tube containing 5.0 ml sterile saline

5-ml pipette

Oxidase reagent

Mineral oil

Sterile Pasteur pipette

Culture

Unknown enteric # _____

Techniques Required

Exercises 10, 13, 14, 15, 16, and 17.

Procedure

Isolation
1. Streak the nutrient agar plate with your unknown for isolation and to determine purity of the culture. Incubate inverted at 35°C for 24 to 48 hours. Record the appearance of the colonies.
2. Determine the oxidase reaction of one of the colonies remaining on the plate. Why? _____

    _____

    How will you determine the oxidase reaction? ___
    _____
    _____

IMViC Tests
1. Inoculate tubes of tryptone broth, MRVP broth, and Simmons citrate agar with your unknown.
2. Incubate the tubes at 35°C for 24 to 48 hours; perform the appropriate tests, and record your results.

Enterotube® (Figure 18.1)
1. Remove both caps from the Enterotube®. One end of the wire is straight and is used to pick up the inoculum; the bent end of the wire is the handle. Holding the Enterotube®, pick a well-isolated col-

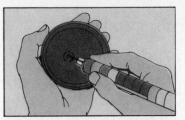

(a) Pick a well-isolated colony with the inoculating end of the wire.

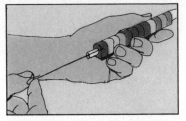

(b) Hold the bent end of the wire and withdraw the needle through all twelve compartments with a turning motion.

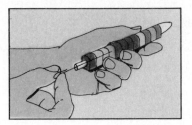

(c) Reinsert the wire through all twelve compartments. Then withdraw to the H₂S/indole compartment. Break the wire at the notch. Replace the caps loosely.

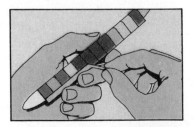

(d) Strip off the blue tape, and slide the clear band over the glucose compartment.

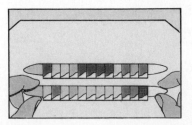

(e) After incubation, compare the tube to an uninoculated one to record results.

Figure 18.1 _____
Inoculating an Enterotube®.

Table 18.2
Enterotube® Biochemical Reactions

| Test | Comments | Indicator Changed | | Refer to Exercise |
		From	To	
GLU	Acid from glucose	Red	Yellow	14
GAS	Gas produced from fermentation of glucose trapped in this compartment, causing separation of the wax			
LYS	Lysine decarboxylase	Yellow	Purple	16
ORN	Ornithine decarboxylase	Yellow	Purple	16
H_2S	Ferrous ion reacts with sulfide ions forming a black precipitate			16
IND	Kovacs reagent is added to the H_2S/IND compartment to detect indole	Beige	Red	16
ADON	Adonitol fermentation	Red	Yellow	14
LAC	Lactose fermentation	Red	Yellow	14
ARAB	Arabinose fermentation	Red	Yellow	14
SORB	Sorbitol fermentation	Red	Yellow	14
V–P	Voges–Proskauer reagents detect acetoin	Beige	Red	14
DUL	Dulcitol fermentation	Green	Yellow	14
PA	Phenylpyruvic acid released from phenylalanine after its deamination combines with iron salts to form a black precipitate			16
UREA	Ammonia changes the pH of the medium	Yellow	Pink	15
CIT	Citric acid used as a carbon source	Green	Blue	18

Source: Roche Diagnostics, Division of Hoffman-LaRoche Inc.

ony with the inoculating end of the wire (Figure 18.1*a*). Avoid touching the agar with the needle.

2. Inoculate the Enterotube® by holding the bent end of the wire and twisting; then withdraw the needle through all twelve compartments using a turning motion (Figure 18.1*b*).

3. Reinsert the needle into the Enterotube®, using a turning motion, through all twelve compartments. Then withdraw to the H_2S/indole compartment. Break the needle at the notch by bending.

> 🔆 **Discard the handle in disinfectant.**

Replace the caps loosely on both ends of the tube (Figure 18.1*c*). The portion of the needle remaining in the tube maintains anaerobic conditions necessary for fermentation, production of gas, and decarboxylation. The part of the wire in the H_2S/

indole compartment will not interfere with these tests.

4. Strip off the blue tape after inoculation but before incubation, to provide aerobic conditions. Slide the clear band over the dextrose (glucose) compartment to contain any small amount of sterile wax that may escape due to excessive gas production by some bacteria (Figure 18.1*d*).

5. Incubate the tube lying on its flat surface at 35°C for 24 hours.

6. Interpret and record all reactions (see Table 18.2) in the Laboratory Report. Read all the other tests before the indole and V–P tests, which follow:

Indole test. Melt a small hole in the plastic film covering the H_2S/indole compartment using a warm inoculating loop. Add 1 to 2 drops of Kovacs reagent, and allow the reagent to contact the agar surface. A positive test is indicated by a red color within 10 seconds.

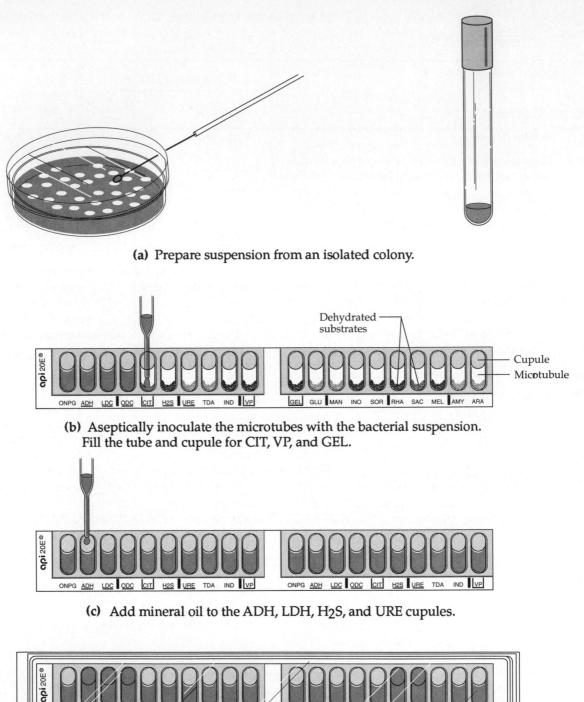

(a) Prepare suspension from an isolated colony.

(b) Aseptically inoculate the microtubes with the bacterial suspension. Fill the tube and cupule for CIT, VP, and GEL.

(c) Add mineral oil to the ADH, LDH, H$_2$S, and URE cupules.

(d) Incubate the strip in its plastic tray.

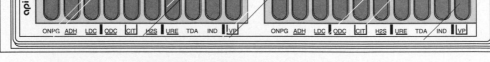

Figure 18.2

Inoculating the API 20E® system.

V–P test. Add 2 drops of 20% KOH containing 5% α-naphthol to the V–P compartment. A positive test is indicated by development of a red color within 20 minutes.

7. Indicate each positive reaction by circling the number appearing below the appropriate compartment of the Enterotube® outlined in the Laboratory Report. Add the circled numbers only within each bracketed section and enter this sum in the space provided below the arrow. Read the five numbers in these spaces across as a five-digit number in the *Computer Coding and Identification System.**

8. Dispose of the Enterotube® by placing in the autoclave basket.

API 20E® (Figure 18.2)

1. Prepare a bacterial suspension by touching the center of a well-isolated colony with a sterile loop, and thoroughly mix the inoculum in sterile saline (Figure 18.2*a*).

2. Place 5 ml of tap water into the corrugated incubation tray, to provide a humid atmosphere during incubation.

3. Using a sterile Pasteur pipette, tilt the API 20E® tray and fill the tube section of the microtubes with the bacterial suspension. Fill the tube *and* cupule sections of the CIT, VP, and GEL tubes (Figure 18.2*b*).

4. After inoculation, completely fill the cupule section of the ADH, LDC, ODC, H₂S, and URE tubes with mineral oil to create anaerobic conditions (Figure 18.2*c*).

5. Place the plastic lid on the tray and incubate the strip for 24 hours at 35°C (Figure 18.2*d*). If the strip

cannot be read after 24 hours, remove the strips from the incubator and refrigerate.

6. Interpret and record all reactions (see Table 18.3) in the Laboratory Report. Read all the other tests before the TDA, VP, and IND, which follow:

TDA test. Add 1 drop 10% ferric chloride. A positive test is brown-red. Indole-positive organisms may produce an orange color; this is a negative TDA reaction.

V–P test. Add 1 drop of V–P reagent II, then 1 drop of V–P reagent I. A positive reaction produces a red color (not pale pink) after 10 minutes.

Indole test. Add 1 drop of Kovacs reagent. A red ring after 2 minutes indicates a positive reaction.

Nitrate reduction. Before adding reagents, look for bubbles in the GLU tube. Bubbles indicate reduction of nitrate to N₂. Add 2 drops of nitrate reagent A and 2 drops of nitrate reagent B. A positive reaction (red) may take 2–3 minutes to develop. A negative test can be confirmed with zinc dust (Exercise 17).

7. Indicate each positive reaction with a + in the appropriate compartment of the Laboratory Report. Add the points for each positive reaction within each bold-outlined section. Read the seven numbers across as a seven-digit number in the *API 20E Analytical Profile Index.** Nitrate reduction is a confirming test and not part of the seven-digit code.

8. Dispose of the API strip, tray, and lid by placing in the autoclave basket.

*Roche Diagnostics, Division of Hoffman-LaRoche Inc., Nutley, NJ 07110.

*API Analytab Products, Division of Sherwood Medical, Plainview, NY 11803.

Table 18.3

API 20E® Biochemical Reactions

Test	Comments	Indicator		Refer to Exercise
		Positive	Negative	
ONPG	O-nitrophenyl-β-D-galactopyranoside is hydrolyzed by the enzyme that hydrolyzes lactose	Yellow	Colorless	14
ADH	Arginine dihydrolase transforms arginine into ornithine, NH_3, and CO_2	Red	Yellow	
LDC	Decarboxylation of lysine liberates cadaverine	Red	Yellow	16
ODC	Decarboxylation of ornithine produces putrescine	Red	Yellow	16
CIT	Citric acid used as sole carbon source	Dark blue	Light green	18
H_2S	Blackening indicates reduction of thiosulfate to H_2S	Black	No blackening	16
URE	Urea is hydrolyzed by the enzyme urease to NH_3 and CO_2	Red	Yellow	15
TDA	Deamination of tryptophan produces indole and pyruvic acid	Brown	Yellow	16
IND	Kovacs reagent is added to detect indole	Red ring	Yellow	16
VP	Addition of KOH and α-naphthol detects the presence of acetoin	Red	Colorless	14
GEL	Gelatin hydrolysis	Diffusion of pigment	No diffusion	15
GLU	Fermentation of glucose			
MAN	Fermentation of mannitol			
INO	Fermentation of inositol			
SOR	Fermentation of sorbitol	Yellow or yellow-green	Blue or green	14
RHA	Fermentation of rhamnose			
SAC	Fermentation of sucrose			
MEL	Fermentation of melibiose			
AMY	Fermentation of amygdalin			
ARA	Fermentation of arabinose			
NO_2 N_2 gas	Nitrate reduction	Red Bubbles Yellow after addition of zinc	Yellow No bubbles Orange after reagents and zinc	17

Source: API Analytab Products, Division of Sherwood Medical, Plainview, NY 11803.

EXERCISE 18

Rapid Identification Methods

Name _____

Date _____

Lab Section _____

Purpose _____

_____

_____

Data

Unknown # _____

Appearance on nutrient agar: _____

Oxidase reaction: _____

IMViC

Indicate positive (+) and negative (−) results for each test.

Indole: _____

Methyl red: _____

V–P: _____

Citrate: _____

Enterotube®

Circle the number corresponding to each positive reaction below the appropriate compartment. Then determine the five-digit code.

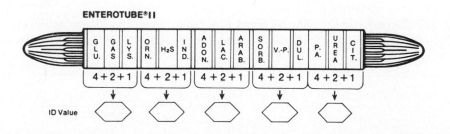

API 20E®

Indicate positive (+) and negative (−) results in the Results line. Then determine the seven-digit code.

	ONPG 1	ADH 2	LDC 4	ODC 1	CIT 2	H₂S 4	URE 1	TDA 2	IND 4	VP 1	GEL 2	GLU 4
Results												
Profile number												

	MAN 1	INO 2	SOR 4	RHA 1	SAC 2	MEL 4	AMY 1	ARA 2	Oxi-dase 4	NO₂	N₂ GAS
Results											
Profile number											

Questions

1. What species was identified in unknown # ———— by the IMViC tests? ————————————

   By the API 20E®? ——————————————————————————————————

   By the Enterotube®? ——————————————————————————————————

2. Did all three methods agree? ———— If not, explain any discrepancies. ——————————
   ——

3. Which method did you prefer? ———————————————— Why? ——————————
   ——

4. Why are systems developed to identify enterics? ——————————————————————
   ——

5. Use Table 18.1 to give an example of a limitation of the IMViC tests. ——————————————
   ——
   ——

6. Why can one species have two or more numbers in a rapid identification system? (For example, *E. coli* is Enterotube® numbers 41310 and 41340; *Citrobacter freundii* is API 20E® numbers 3504552 and 3504553.)
   ——
   ——

7. Why should the first digit in the five-digit Enterotube® code number be equal to or greater than 4? ————

8. Why is an oxidase test performed on a culture before using API 20E® and Enterotube® to identify the culture?
   ——

PART 5

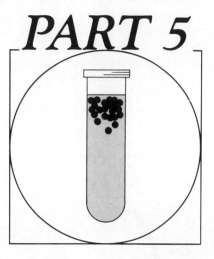

Microbial Growth

EXERCISES

In previous exercises, we studied bacterial growth by observing colonies or turbidity, making visual determinations on an all-or-none basis with the naked eye. In Exercise 20, we will use instrumentation to detect the slight changes in bacterial numbers that occur over time — changes that are not visible to the naked eye — to measure lag and log phases of growth (see the figure).

Bacteria have nutritional, physical, chemical, and environmental requirements that must be met in order for growth to occur (Parts 3 and 4). A knowledge of the conditions necessary for microbial growth can facilitate culturing microorganisms and controlling unwanted organisms.

In 1861, Jean Baptiste Dumas presented a paper to the Academy of Sciences on behalf of Pasteur. In this paper, he stated:

> *The existence of infusoria having the characteristics of ferments is already a fact which seems to deserve attention, but a characteristic which is even more interesting is that these infusoria-animalcules live and multiply indefinitely in absence of the smallest amount of air or free oxygen. . . .*

> *This is, I believe, the first known example of animal ferments and also of animals living without free oxygen gas.**

When we speak of *air* as a growth requirement, we are referring to the oxygen in air. Furthermore, when we refer to oxygen, we usually mean the **molecular oxygen (O_2)** that acts as an electron acceptor in aerobic respiration (Exercise 17). The significance of the presence or absence of oxygen is investigated in Exercise 19.

Physical environmental conditions that influence microbial growth are temperature (Exercise 20), osmotic pressure (Exercise 21), and pH (Exercise 21).

*In H. A. Lechevalier and M. Solotorovsky. *Three Centuries of Microbiology*. New York: Dover Publications, 1974, p. 45.

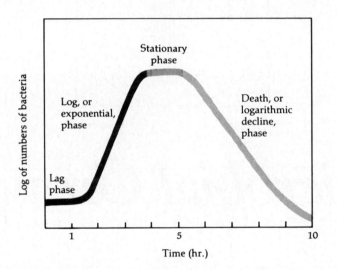

Bacterial growth curve, showing the four typical phases of growth.

Oxygen and the Growth of Bacteria

What is education but a process by which a person begins to learn how to learn?

PETER USTINOV

Objectives

After completing this exercise you should be able to:

1. Provide the incubation conditions for each of the following types of organisms: aerobes, obligate anaerobes, aerotolerant anaerobes, microaerophiles, and facultative anaerobes.
2. Describe three methods of culturing anaerobes.
3. Cultivate anaerobic bacteria.

Background

The presence or absence of molecular oxygen (O_2) can be very important to the growth of bacteria. Some bacteria, called obligate **aerobic bacteria,** require oxygen, while others, called **anaerobic bacteria,** do not use oxygen. One reason **obligate anaerobes** cannot tolerate the presence of oxygen is that they lack catalase, and the resultant accumulation of hydrogen peroxide is lethal (Exercise 17). **Aerotolerant anaerobes** cannot use oxygen but tolerate it fairly well, although their growth may be enhanced by microaerophilic conditions. Most of these bacteria use fermentative metabolism (Exercise 13).

Some bacteria, the **microaerophiles,** grow best in an atmosphere with increased carbon dioxide (7% to 10%) and lower concentrations of oxygen. Microaerophiles will grow in a solid nutrient medium at a depth to which small amounts of oxygen have diffused into the medium (Figure 19.1*e*). In order to culture microaerophiles on Petri plates and nonreducing media, a **candle jar** is used. Inoculated plates and tubes are placed in a large jar with a lighted candle. After the lid is placed on the jar, the candle will extinguish when the concentration of oxygen decreases (the concentration of carbon dioxide will be raised).

The majority of bacteria are capable of living with or without oxygen; these bacteria are called **facultative anaerobes** (Figure 19.1*b*).

Four genera of bacteria lacking catalase are *Streptococcus, Leuconostoc, Lactobacillus,* and *Clostridium.* Species of *Clostridium* are obligate anaerobes, but the other three are aerotolerant anaerobes. The three aerotolerant anaerobes lack the cytochrome system to produce hydrogen peroxide and therefore do not need catalase. Determining the presence or absence of catalase can be very helpful in identifying bacteria. When a few drops of 3% hydrogen peroxide are added to a

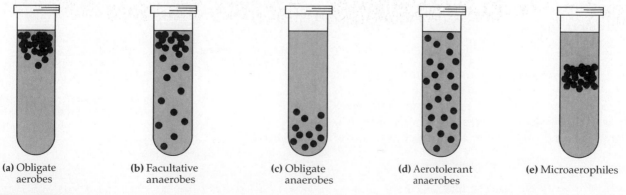

(a) Obligate aerobes   **(b)** Facultative anaerobes   **(c)** Obligate anaerobes   **(d)** Aerotolerant anaerobes   **(e)** Microaerophiles

Figure 19.1 _____

Effect of oxygen concentration on the growth of various types of bacteria in a tube of solid medium. Oxygen diffuses only a limited distance from the atmosphere into the solid medium.

microbial colony and catalase is present, molecular oxygen is released as bubbles.

In the laboratory, we can culture anaerobes either by excluding free oxygen from the environment or by using reducing media. Many anaerobic culture methods involve both processes. Some of the anaerobic culturing methods are as follows:

1. **Reducing media** contain reagents that chemically combine with free oxygen, reducing the concentration of oxygen. In thioglycollate broth, **sodium thioglycollate** ($HSCH_2COONa$) will combine with oxygen. A small amount of agar is added to increase the viscosity, and this reduces the diffusion of air into the medium. Usually dye is added to indicate where oxygen is present in the medium. Resazurin, which is pink in the presence of excess oxygen and colorless when reduced, or methylene blue (see below) are commonly used indicators.

2. Conventional, nonreducing media can be incubated in an anaerobic environment. Oxygen is excluded from a **Brewer anaerobic jar** by adding a Gas-Pak® and a palladium catalyst to catalytically combine the hydrogen with oxygen and form water (Figure 19.2). Carbon dioxide and hydrogen are given off when water is added to the Gas-Pak® envelope of sodium bicarbonate and sodium borohydride. A methylene blue indicator strip is placed in the jar; methylene blue is blue in the presence of oxygen and white when reduced. One of the disadvantages of the Brewer jar is that the jar must be opened to observe or use one plate. An inexpensive modification of the Brewer jar has been developed using disposable plastic bags for one or two culture plates (Figure 19.3). In this technique, the bag is coated with antifogging chemicals, and a wet sodium bicarbonate tablet is added to generate carbon dioxide. Iron (Fe^0) activated with water removes O_2 as iron oxide (Fe_2O_3) is produced. One plate can be observed without opening the bag.

3. **Anaerobic incubators** and **glove boxes** can also be used for incubation. Air is evacuated from the chamber and can be replaced with a mixture of carbon dioxide and nitrogen.

All methods of anaerobic culturing are only effective if the specimen or culture of anaerobic organisms is collected and transferred in a manner that minimizes exposure to oxygen. In this exercise, we will try two methods of anaerobic culture and will perform the catalase test.

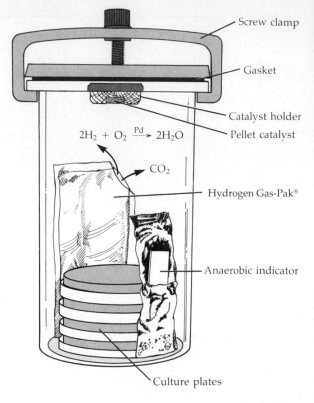

Screw clamp

Gasket

Catalyst holder

Pellet catalyst

$2H_2 + O_2 \xrightarrow{Pd} 2H_2O$

CO_2

Hydrogen Gas-Pak®

Anaerobic indicator

Culture plates

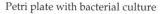

Figure 19.2

Brewer anaerobic jar. Carbon dioxide and hydrogen are generated by the addition of water to a chemical packet. Hydrogen gas combines with atmospheric oxygen in the presence of a palladium catalyst (in the lid or in the Gas-Pak® Plus packet) to form H_2O. The anaerobic indicator contains methylene blue, which is blue when oxidized and turns colorless when the oxygen is removed from the container.

Petri plate with bacterial culture

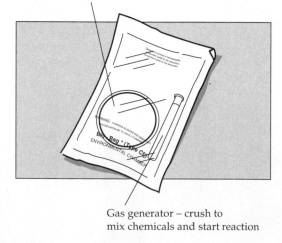

Gas generator – crush to mix chemicals and start reaction

Figure 19.3

Plastic bag with Petri plate and CO_2 generator.

Materials

Petri plates containing phenylethylalcohol agar (PEA) (2)

Petri plates containing blood agar (2)

Tube containing thioglycollate broth with indicator

Tube containing sterile cotton swab

Brewer anaerobic jars per lab section (2)

3% hydrogen peroxide (H_2O_2)

Soil slurry

Techniques Required

Exercises 10, 11, 12, and 17. Read about blood agar (Exercise 46).

Procedure

1. Don't shake the thioglycollate. Why? _____

    _____

    What should you do to salvage it if you do shake it?

    _____

    Aseptically inoculate the thioglycollate with a loopful of the soil slurry. Incubate at room temperature until good growth is seen (2 days to 1 week). Describe the oxygen requirements of the resulting growth.

2. Dip the cotton swab in the soil slurry, and swab one-half of each plate: two PEA and two blood agar plates.

> ☣ **Discard the swab in disinfectant.**

3. Using a sterile loop, streak back and forth into the swabbed area a few times, then streak away from the inoculum (Figure 19.4).

4. Label one PEA and one blood agar "Aerobic" and incubate them inverted at 22°C for 2 days. Label the

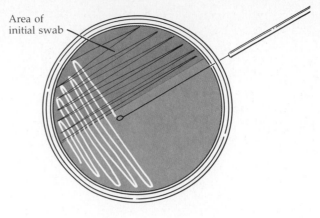

Area of initial swab

Figure 19.4 _____

Using a sterile loop, spread the inoculum over the remaining half of the plate.

other two plates "Anaerobic" and place them, inverted, in the Brewer jar.

5. Your instructor will demonstrate how the jar is rendered anaerobic once filled with plates. Ten ml of water are added to a Gas-Pak® generator and placed in the jar with a methylene blue indicator. What causes the condensation that forms on the sides of the jar? _____

    _____

6. Incubate the jar for 2 days at 22°C.

7. After incubation, describe the colonies on each plate. Look for hemolysis (Exercise 46) on the blood agar.

8. Perform the catalase test by adding a few drops of 3% H_2O_2 to the different colonies on the PEA media. A positive catalase test produces a bubbling white froth. A dissecting microscope (Appendix E) can be used if more magnification is required to detect bubbling. The catalase test may also be done by transferring bacteria to a slide and adding the H_2O_2 to it. Bacterial growth on blood agar must be tested this way. Why? _____

    _____

EXERCISE 19

Oxygen and the Growth of Bacteria

Name _____

Date _____

Lab Section _____

Purpose _____

_____

_____

Data

Thioglycollate

Sketch the location of bacterial growth, as in Figure 19.1, and label as to its oxygen requirements. Observe other students' thioglycollate tubes.

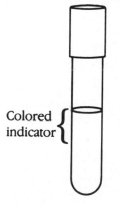

Colored indicator {

PEA Plates

Aerobic Number of different colonies _____		Anaerobic _____	
Colony description	Catalase reaction	Colony description	Catalase reaction

Blood Agar Plates

Aerobic Number of different colonies _____			Anaerobic _____		
Colony description	Hemolysis	Catalase reaction	Colony description	Hemolysis	Catalase reaction

Do the anaerobic plates have an odor? _____

Questions

1. Which of the colonies on the Petri plates could be anaerobic bacteria? _____

   _____

   What would you need to do to determine whether they are anaerobes? _____

   _____

2. Could you have a catalase-positive colony on the plate incubated anaerobically? _____

   _____

3. Why will obligate anaerobes grow in thioglycollate? _____

   _____

   _____

4. The catalase test is often used clinically to distinguish between two genera of gram-positive cocci: _____

   and _____ . It is also used to distinguish between two genera of gram-positive

   rods: _____ and _____ .

5. To what do you attribute the odors of anaerobic decomposition? _____

   _____

   _____

6. The following genera of anaerobes are commonly found in feces. How would you differentiate them? *Bacteroides, Fusobacterium, Clostridium, Enterococcus, Veillonella, Lactobacillus.*

   _____

   _____

EXERCISE 20

Determination of a Bacterial Growth Curve: The Role of Temperature

Objectives

After completing this exercise you should be able to:

1. Identify the four phases of a typical bacterial growth curve.
2. Measure bacterial growth turbidometrically.
3. Interpret growth data plotted on a graph.
4. Determine the effect of temperature on bacterial growth.

Background

The phases of growth of a bacterial population can be determined by measuring the turbidity of a broth culture (see the figure on p. 116). Turbidity is not a direct measure of bacterial numbers, but increasing turbidity does indicate bacterial growth. Since millions of cells per milliliter must be present to discern turbidity with the eye, a spectrophotometer is used to detect the presence of smaller numbers of bacteria (Appendix C).

Most bacteria grow within a particular temperature range (Figure 20.1a). The *minimum growth temperature* is the lowest temperature at which a species will grow. A species grows fastest at its *optimum growth temperature*. And the highest temperature at which a species can grow is its *maximum growth temperature*. Some heat is necessary for growth. Heat probably increases the rate of collisions between molecules, thereby increasing enzyme activity. At temperatures near the maximum growth temperature, growth ceases, presumably due to inactivation of enzymes.

Bacteria can be classified into three groups, based on their growth at various temperatures (Figure 20.1b). The optimum temperature of **psychrophiles** is between 0°C and 20°C. The optimum temperature for **mesophiles** is between 20°C and 45°C. The optimum temperature for many **thermophiles** is between 45°C and 60°C, although some are capable of growth above 90°C. The range of temperature preferred by bacteria is genetically determined, resulting in enzymes with different temperature requirements.

In this exercise, we will plot growth curves for a bacterium at different temperatures to determine the preferred temperature range of this species.

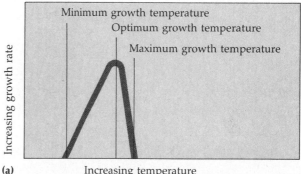

(a)   Increasing temperature

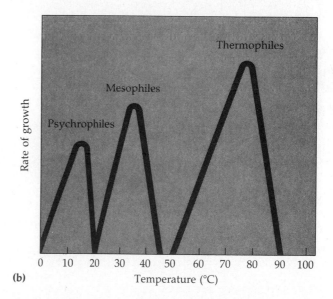

(b)

Figure 20.1
(a) Growth response of bacteria within their growing temperature range. **(b)** Typical growth responses to temperature for psychrophiles, mesophiles, and thermophiles.

Materials

Flask containing nutrient broth + 1.5% NaCl

Spectrophotometer tubes

Sterile 5-ml pipettes (11)

Spectrophotometer

123

Culture

Vibrio natriegens

Techniques Required

Exercise 10 and Appendices A, C, and D.

Procedure

1. Aseptically transfer 4 ml nutrient broth to one of the spectrophotometer tubes. This is your control to *standardize* the spectrophotometer.
2. Aseptically inoculate a flask of nutrient broth with 5 ml *Vibrio*.
3. Swirl to mix the contents. Transfer 4 ml from the flask to the second spectrophotometer tube, and measure the absorbance (Abs.) on the spectro-

photometer. *Read* Appendix C before using the spectrophotometer.

4. Record all measurements.

> **After taking a reading, discard the 4 ml into disinfectant.**

5. Place the flask at your *assigned* temperature.
6. Record the absorbance every 10 minutes for 60 to 90 minutes. Remove the flask from the water bath for the minimum time possible to take each sample. To take a sample, aseptically pipette 4 ml into the second spectrophotometer tube.
7. Graph the data you obtained. *Read* Appendix D before drawing your graph.

EXERCISE 20

Determination of a Bacterial Growth Curve: The Role of Temperature

Name _____

Date _____

Lab Section _____

Purpose _____

_____

_____

Data

Temperature: _____

Time (min)	Absorbance	Time (min)	Absorbance
0			

Plot your data on the semi-log graph paper. Record absorbance on the Y-axis (logarithmic scale) and time on the X-axis (arithmetic scale). Obtain data for other temperatures from your classmates. Plot these data on the same graph using different lines or colors. Label the phases of growth.

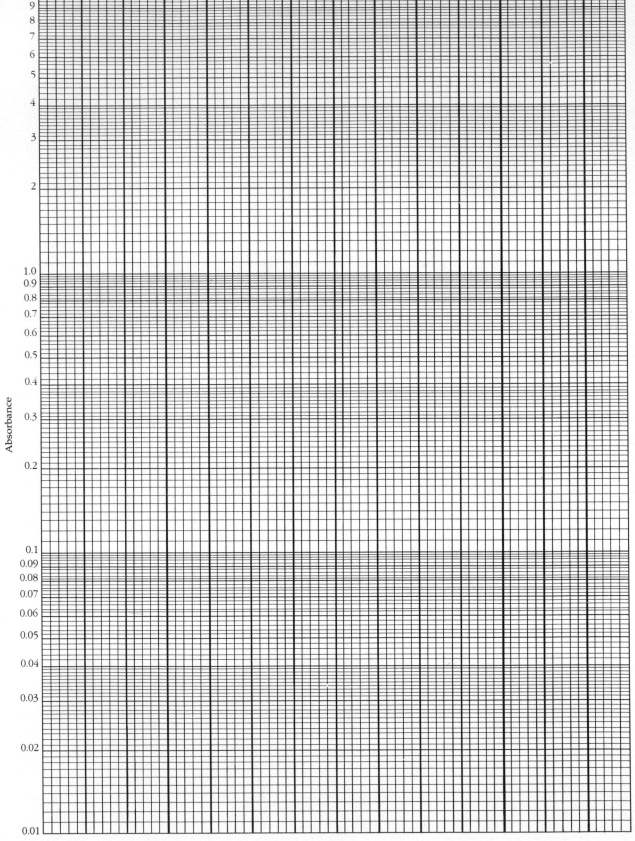

Absorbance

Time

Conclusions

1. Summarize the effect of temperature on growth of *Vibrio*. _____

    _____

    _____

    _____

2. What is *Vibrio's* optimum temperature? _____

Questions

1. Why aren't you likely to get lag and death phases in this experiment? _____

    _____

    _____

2. How could you use turbidity to estimate the numbers of bacteria? _____

    _____

    _____

3. What is the optimum temperature for human pathogens? _____

4. Where would you most likely find a thermophile? _____

    A psychrophile? _____

5. What is the effect of temperature on enzymes? Can you use any examples from your experiments? _____

    _____

    _____

Other Influences on Microbial Growth: Osmotic Pressure and pH

Objectives

After completing this exercise you should be able to:

1. Define osmotic pressure and explain how it affects a cell.
2. Explain how microbial growth is related to osmotic pressure and pH.
3. Prepare and use a gradient plate.

Background

The osmotic pressure of an environment influences microbial growth. **Osmotic pressure** is the force with which a solvent moves from a solution of lower solute concentration to a solution of higher solute concentration across a semipermeable membrane (Figure 21.1). The addition of solutes, especially salts and sugars, and the resultant increase in osmotic pressure, are used to preserve some foods. Can you name a food that is preserved with salt? _____
With sugar? _____

Bacteria are often able to live in a **hypotonic** environment in which the concentration of solutes outside the cell is lower than inside the cell. However, when the concentration of solutes outside the cell is higher than that inside the cell—a **hypertonic** environment—a bacterial cell will undergo plasmolysis. *Plasmolysis* occurs when water leaves the cell and the cytoplasmic membrane draws inward, away from the cell wall. A few bacteria called **facultative halophiles,** are able to tolerate salt concentrations up to 10%, and **extreme halophiles** require 15% to 20% salt.

The acidity or alkalinity **(pH)** of the environment also influences microbial growth. Optimal bacterial growth usually occurs between pH 6.5 and pH 7.5. Only a few bacteria grow at an acidic pH below 4.0. Therefore, organic acids are used to preserve foods such as fermented dairy products (Exercise 54).

Many bacteria produce acids that may inhibit their growth. **Buffers** are added to culture media to neutralize these acids. The peptones in complex media act as buffers. Phosphate salts are often used as buffers in chemically defined media.

In this exercise, we will compare bacterial growth to fungal growth. Fungi differ from bacteria morphologically and biochemically. The yeast (Exercise 33) and

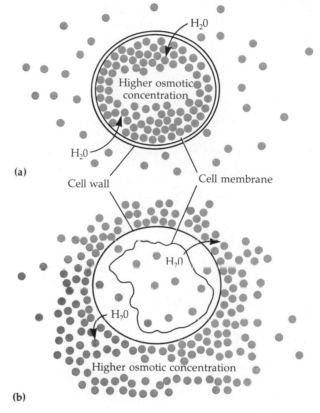

(a)

(b)

Figure 21.1 _____

Osmotic pressure. **(a)** Water moves into the cell in a hypotonic solution. **(b)** Water leaves the cell in a hypertonic environment.

mold (Exercise 34) used here are microbes belonging to the Kingdom Fungi.

Materials

Effect of Osmotic Pressure on Growth

(Your instructor will assign salt or sugar.)

Petri plates containing nutrient agar + 5%, 10%, 15% NaCl (one of each) *or*

Petri plates containing nutrient agar + 10%, 25%, 50% sucrose (one of each) *or*

Petri plate containing nutrient agar

Gradient Plate

Sterile Petri plate

Tube containing melted nutrient agar

Tube containing melted nutrient agar + 25% NaCl *or* 50% sucrose

Salt *or* sugar enrichments

Effect of pH on Growth

Tubes containing nutrient broth adjusted to pH 2.5, 5.0, 7.0, 9.5 (one of each)

Cultures

Effect of Osmotic Pressure on Growth

Escherichia coli

Staphylococcus aureus

A yeast *(Saccharomyces)*

A mold *(Penicillium)*

Effect of pH on Growth (as assigned)

Staphylococcus aureus

Alcaligenes faecalis

Escherichia coli

Serratia marcescens

A mold *(Penicillium)*

A yeast *(Saccharomyces)*

Techniques Required

Exercises 10 and 11.

Procedure

Effect of Osmotic Pressure on Growth

1. Obtain a set of nutrient agar + salt *or* nutrient agar + sucrose plates. Draw two crossed lines to form four quadrants on the bottom of one plate. Each quadrant will be inoculated with one of the organisms and should be labeled accordingly (Figure 21.2). Repeat this procedure until all plates are marked. One student group will mark the *control* plate (nutrient agar alone). What is the purpose of the control plate? _____

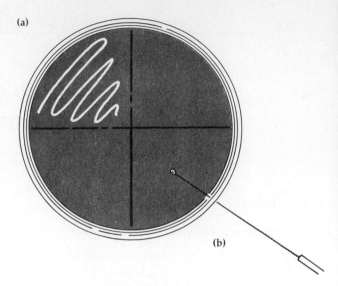

(a)

(b)

Figure 21.2 _____

Inoculate each plate with four organisms. **(a)** Streak bacteria and yeast cultures. **(b)** Place a loopful of mold suspension in the center of one quadrant.

2. Using a sterile loop, inoculate *S. aureus* onto one quadrant on each plate (Figure 21.2*a*). Repeat this procedure to inoculate *E. coli* and the yeast.
3. Place a loopful of the mold suspension in the center of the remaining quadrant on each plate (Figure 21.2*b*).
4. Invert the plates and incubate at 35°C for 24 to 48 hours. Record the relative amounts of bacterial growth; the culture showing the "best" growth is given (+4) and others are evaluated relative to that culture.
5. Incubate the plate at room temperature for 2 to 4 days, and record relative amounts of mold growth.

Gradient Plate

1. Pour a layer of nutrient agar into a sterile Petri plate and let it solidify, with the plate resting on a small pencil or a loop handle (Figure 21.3*a*) so that a wedge forms.
2. On the bottom of the plate, draw a line at the end corresponding to the high side of the agar and label it "low." Draw four lines perpendicular to the first line, and, if assigned sucrose, label the lines "0.5," "5," "10," and "15"; if assigned salt, "0.5," "5," "15," and "30" (Figure 21.3*b*).
3. Pour a layer of salt *or* sugar agar over the nutrient agar wedge. Let it solidify. A solute gradient concentration has been prepared. Why was the high side of the agar labeled "low" in step 2? _____

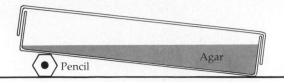

(a) Allow the nutrient agar to solidify into a wedge.

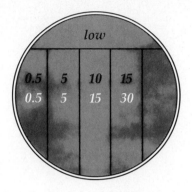

(b) Mark and label the plate.

Figure 21.3 _____

Gradient plate.

4. The following enrichments (Exercise 12) have been prepared:
   a. Raisins were placed in tubes of nutrient broth + 0.5% sucrose, 5% sucrose, 10% sucrose, and 15% sucrose.
   b. Hamburger was inoculated into tubes of nutrient broth + 0.5% NaCl, 5% NaCl, 15% NaCl, and 30% NaCl.
5. Using the enrichment with the same solute as your gradient plate, streak a loopful of each enrichment onto the agar over the prelabeled lines. Streak from the low to the high end.
6. Incubate the plate inverted at 35°C for 48 hours, and record the growth patterns.

Effect of pH on Growth

1. Inoculate a loopful of your organism into each tube of pH test broth.
2. Incubate the tubes at 35°C for 24 to 48 hours. Record your results and results for the organisms tested by other students.

EXERCISE 21

Other Influences on Microbial Growth: Osmotic Pressure and pH

Name _____

Date _____

Lab Section _____

Purpose _____

_____

_____

Data

Effect of Osmotic Pressure on Growth

Rate the relative amounts of growth on nutrient agar and nutrient agar-containing solutes (salt and sucrose): (−) = no growth; (+) = minimal growth; (2+) = moderate growth; (3+) = heavy growth; (4+) = very heavy (maximum) growth.

Medium	Organism/Amount of Growth			
	S. aureus	E. coli	Yeast	Mold
Nutrient agar				
+ 5% NaCl				
+10% NaCl				
+15% NaCl				
+10% sucrose				
+25% sucrose				
+50% sucrose				

Gradient Plate

Sketch your growth patterns and those of another group.

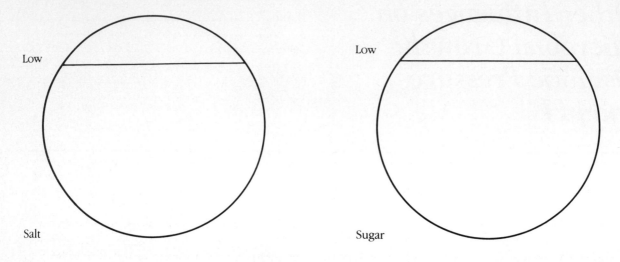

Effect of pH on Growth

Record the growth (+) or lack of growth (−) of each organism used.

pH	Growth of					
	S. aureus	*A. faecalis*	*E. coli*	*S. marcescens*	Mold	Yeast
2.5						
5.0						
7.0						
9.5						

Questions

1. Which organism tolerates high concentrations of salt best? _____

   Of what advantage is this to the organism? _____

   _____

2. Which organism tolerates high concentrations of sucrose best? _____

   Why is this advantageous to the organism? _____

   _____

3. Write a statement comparing the osmotic tolerance of bacteria to that of fungi. _____

_____

4. Explain the growth patterns on the gradient plates. _____

_____

_____

_____

_____

5. What is the pH tolerance of bacteria compared to that of fungi? _____

_____

6. Did any organisms grow in acidic conditions? _____ Why would you expect some

organisms to grow in an acidic environment? _____

_____

_____

_____

7. Why are bacteria more likely to survive in a hypotonic environment than in a hypertonic environment?

_____

8. What is the principle of the gradient plate? _____

_____

9. What is the purpose of the control plate? _____

_____

10. Should (4 +) growth occur only on the control plate? _____

Briefly explain. _____

_____

11. Why were the media containing raisins and hamburger labeled "enrichments"? _____

_____

12. What is meant by "neutral pH"? _____

_____

PART 6

Control of Microbial Growth

The destruction of microbes by heat was employed by Lazzaro Spallanzani in the 1760s. He heated nutrient broth to kill preexisting life in his attempts to disprove the concept of spontaneous generation. In the 1860s, Pasteur heated broth in specially designed flasks and ended the debate over spontaneous generation (see the figure). In Exercise 22, we will examine the effectiveness of heat for killing microbes.

The surgeon Joseph Lister was greatly influenced by Pasteur's demonstrations of the omnipresence of microorganisms and his proof that microorganisms cause decomposition of organic matter. Lister had observed the disastrous consequences of compound bone fractures (in which the skin is broken) compared to the relative safety of simple bone fractures. He had heard of the treatment of sewage with carbolic acid (phenol) to prevent diseases of cattle that had come into contact with the sewage. In 1867, Lister wrote: "It appears that all that is requisite is to dress the wound with some material capable of killing these septic germs"* to prevent disease and death due to microbial growth.

In Part 6, we will examine current methods of controlling microbial growth, with special attention on the use of disinfectants and antiseptics (Exercise 24) and hand scrubbing (Exercise 26).

*Quoted in W. Bulloch. *The History of Bacteriology.* New York: Dover Publications, 1974, p. 46.

Pasteur's experiment that disproved spontaneous generation involved boiling nutrient broth in an S-neck flask. Microorganisms did not appear in the cooled broth even after a long period of time.

Physical Methods of Control: Heat

The successful man lengthens his stride when he discovers that the sign post has deceived him; the failure looks for a place to sit down.

JOHN R. ROGERS

Objectives

After completing this exercise you should be able to:

1. Compare the bactericidal effectiveness of dry heat and moist heat on different species of bacteria.
2. Evaluate the heat tolerance of microbes.
3. Define and provide a use for each of the following: incineration, hot air oven, pasteurization, boiling, and autoclaving.

Background

The use of extreme temperature to control the growth of microbes is widely employed. Generally, if heat is applied, bacteria are killed; if cold temperatures are utilized, bacterial growth is inhibited.

Heat sensitivity of organisms can be affected by container size, cell density, moisture content, pH, and medium composition. Bacteria exhibit different tolerances to the application of heat. Heat sensitivity is genetically determined and is partially reflected in the optimal growth ranges (Exercise 20), which are **psychrophilic** (0°C to 20°C), **mesophilic** (20°C to 45°C), and **thermophilic** (45°C to 90°C), and by the presence of heat-resistant endospores (Exercise 7). Overall, bacteria are more heat resistant than most other forms of life.

Heat can be applied as dry or moist heat. **Dry heat,** such as that in hot air ovens or incineration (for example, flaming loops), denatures enzymes, dehydrates microbes, and kills by oxidation effects. A standard application of dry heat in a hot air oven is 170°C for 2 hours. The heat of hot air is not readily transferred to a cooler body such as a microbial cell. Moisture transfers heat energy to the microbial cell more efficiently than dry air, resulting in the denaturation of enzymes, **Moist heat** methods include pasteurization, boiling, and autoclaving. In **pasteurization** the temperature is maintained at 63°C for 30 minutes or 72°C for 15 seconds to kill designated organisms that are pathogenic or cause spoilage. **Boiling** (100°C) for 10 minutes will kill vege-

Table 22.1

Relationship Between Pressure and Temperature of Steam

Pressure (pounds per square inch, psi, in excess of atmospheric pressure)	Temperature (°C)
0 psi	110°C
5 psi	110°C
10 psi	116°C
15 psi	121°C
20 psi	126°C
30 psi	135°C

Source: G. J. Tortora, B. R. Funke, and C. L. Case. *Microbiology: An Introduction,* 4th ed. Redwood City, CA: Benjamin/Cummings, 1992.

tative bacterial cells; however, endospores are not inactivated. The most effective method of moist heat sterilization is **autoclaving,** the use of steam under pressure. Increased pressure raises the boiling point of water and produces steam with a higher temperature (Table 22.1). Standard conditions for autoclaving are 15 psi, 121°C for 15 minutes. This is usually sufficient to kill endospores and render materials sterile.

There are two different methods of measuring heat effectiveness. **Thermal death time (TDT)** is the length of time required to kill all bacteria in a liquid culture at a given temperature. The less common **thermal death point (TDP)** is the temperature required to kill all bacteria in a liquid culture in 10 minutes.

Materials

Petri plates containing nutrient agar (2)

Thermometer

Empty tube

Beaker

Hot plate or tripod and asbestos pad

Ice

Cultures (as assigned)

Group A:
    Old (48 to 72 hours) *Bacillus subtilis*
    Young (24 hours) *Bacillus subtilis*

Group B:
    Staphylococcus epidermidis
    Escherichia coli

Group C:
    Young (24 hours) *Bacillus subtilis*
    Escherichia coli

Group D:
    Mold *(Penicillium)* spore suspension
    Old (48 to 72 hours) *Bacillus subtilis*

Demonstration
Autoclave and dry heat methods

Techniques Required

Exercises 10 and 11.

Procedure

Each pair of students is assigned two cultures and a temperature.

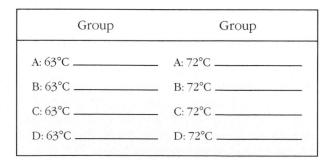

Group	Group
A: 63°C _____	A: 72°C _____
B: 63°C _____	B: 72°C _____
C: 63°C _____	C: 72°C _____
D: 63°C _____	D: 72°C _____

You can share beakers of water as long as the same temperature is being evaluated.

1. Divide two plates of nutrient agar into five sections each. Label the sections "0," "15 sec," "2 min," "5 min," and "15 min."

2. Set up a water bath in the beaker, with the water level higher than the level of the broth in the tubes. Do not put the broth tubes into the water bath at this time. Carefully put the thermometer in a test tube of water in the bath.

3. Streak the assigned organisms on the "0" time section of the appropriate plate. Why are we using "old" and "young" *Bacillus* cultures? _____ _____

4. Raise the temperature of the bath to the desired temperature and maintain that temperature. Use ice to adjust the temperature. Why was 63°C selected as one of the temperatures? _____ _____

5. Place the broth tubes of your organism into the bath when the temperature is at the desired point. After 15 seconds, remove the tubes, resuspend the culture, streak a loopful on the corresponding sections, and return the tubes to the water bath. Repeat at 2, 5, and 15 minutes. What is the longest time period that any microbe is exposed to heat? _____ _____

6. When done, clean the beaker and return the materials. Incubate the plates inverted at 35°C until the next lab period. Record your results and the results for the other organisms tested: $(-)$ = no growth; $(+)$ = minimum growth; $(2+)$ = moderate growth; $(3+)$ = heavy growth; and $(4+)$ = maximum growth.

7. Examine the demonstration plates and record your observations.

Physical Methods of Control: Heat

Name _____

Date _____

Lab Section _____

Purpose _____

_____

_____

Data

Record growth on a scale from (−) to (4 +).

Organism	Temperature/Time									
	63°C					72°C				
	0	15 sec	2 min	5 min	15 min	0	15 sec	2 min	5 min	15 min
Old *Bacillus subtilis*										
Young *Bacillus subtilis*										
Staphylococcus epidermidis										
Escherichia coli										
Mold (*Penicillium*) spores										

Demonstration Plates

	Control	Autoclaved	Dry-heated
Number of colonies			
Number of different colonies			

Conclusions

Questions

1. Explain any unexpected results.

2. What is the heat sensitivity of fungal spores in comparison to bacterial endospores?

3. Give an example of an application (use) of thermal death time.

4. In the exercise, was the thermal death time or thermal death point determined?

5. Give an example of a nonlaboratory use of each of the following to control microbial growth: incineration, pasteurization, autoclaving.

6. Define pasteurization. What is the purpose of pasteurization?

7. Compare the effectiveness of autoclaving and dry heat.

Physical Methods of Control: Ultraviolet Radiation

Objectives

After completing this exercise you should be able to:

1. Examine the effects of ultraviolet radiation on bacteria.
2. Explain the method of action of ultraviolet radiation and light repair of mutations.

Background

Radiant energy comes to the earth from the sun and other extraterrestrial sources, and some is generated on earth from natural and man-made sources. The **radiant energy spectrum** is shown in Figure 23.1. Radiation differs in wavelength and energy. The shorter wavelengths have more energy. X rays and gamma rays are forms of **ionizing radiation.** Their principal effect is to ionize water into *highly reactive free radicals* (with unpaired electrons) that can break strands of DNA. The effect of radiation is influenced by many variables, such as the age of the cells, media composition, and temperature.

Some **nonionizing** wavelengths are essential for biochemical processes. The main absorption wavelengths for green algae, green plants, and photosynthetic bacteria are shown in Figure 23.1*a*. Animal cells synthesize vitamin D in the presence of light around 300 nm. Nonionizing wavelengths are harmful to most bacteria. The most lethal wavelength is 265 nm, which corresponds to the optimal absorption wavelength of DNA (Figure 23.1*b*). Ultraviolet light induces *pyrimidine dimers* in the nucleic acid, which result in a mutation. Mutations in critical genes result in the death of the cell unless the damage is repaired. When pyrimidine dimers are exposed to visible light, the enzyme pyrimidine dimerase is activated and splits the dimer. This is called **light repair** or **photoreactivation.** Another repair mechanism, called **dark repair,** is independent of light. Dimers are removed by endonuclease, DNA polymerase replaces the bases, and DNA ligase seals the sugar-phosphate backbone.

As a sterilizing agent, ultraviolet radiation is limited by its poor penetrating ability. It is used to sterilize some heat-labile solutions and to decontaminate hospital operating rooms and food-processing areas.

In this exercise we will investigate the effects of ultraviolet radiation and light repair using lamps of the desired wavelength.

Materials

Petri plates containing nutrient agar (3)

Sterile cotton swabs (3)

Cardboard or prepared templates (13 cm × 13 cm)

Ultraviolet lamp (265 nm)

Scissors

Cultures (as assigned)

Serratia marcescens

Bacillus subtilis

Techniques Required

Exercise 10.

Procedure

1. Swab the surface of each plate with either *Serratia* or *Bacillus;* to ensure complete covering, swab the surface in two directions. Label the plates "A," "B," and "C."
2. Select a cardboard template, or design your own by cutting a pattern in a piece of cardboard with scissors.

> ⚠ **Do not look at the ultraviolet light, and do not leave your hand exposed to it.**

3. Place each plate directly under the ultraviolet light about 5 cm from the light with the *cover off,* agar-side up, and the template over the plate according to the following protocol. Why should the cover be removed? _____

_____

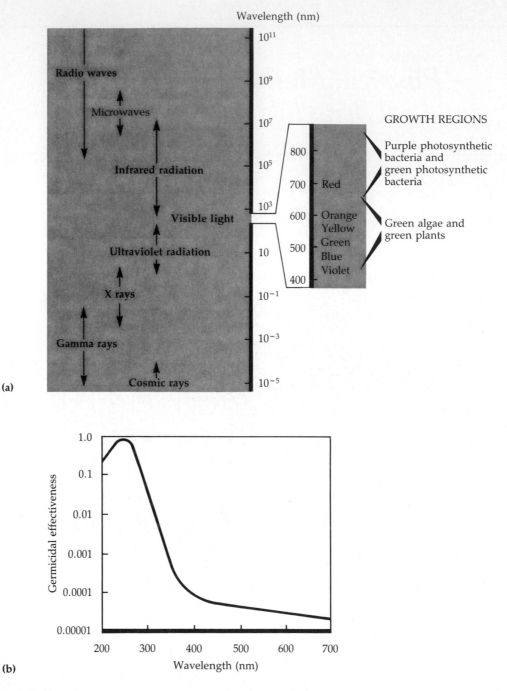

Wavelength (nm)

GROWTH REGIONS

Purple photosynthetic
bacteria and
green photosynthetic
bacteria

Green algae and
green plants

Figure 23.1

Radiant energy. **(a)** Radiant energy spectrum and absorption of light for photosynthesis. **(b)** Germicidal effectiveness of radiant energy between 200 and 700 nanometers (nm) (from UV to visible red light).

Plate A: Expose for 30 seconds and put in a dark incubator (22°C).

Plate B: Expose for 30 seconds and incubate in a room with the lights on.

Plate C: Expose for 90 seconds and put in a dark incubator (22°C).

4. Incubate all three plates inverted at 22°C or at room temperature until the next period.

5. Examine all plates and record your results. Observe the results of students using the other bacteria.

EXERCISE 23

Physical Methods of Control: Ultraviolet Radiation

Name _____

Date _____

Lab Section _____

Purpose _____

_____

_____

Data

Sketch your results and those for the other culture used. Note any pigmentation.

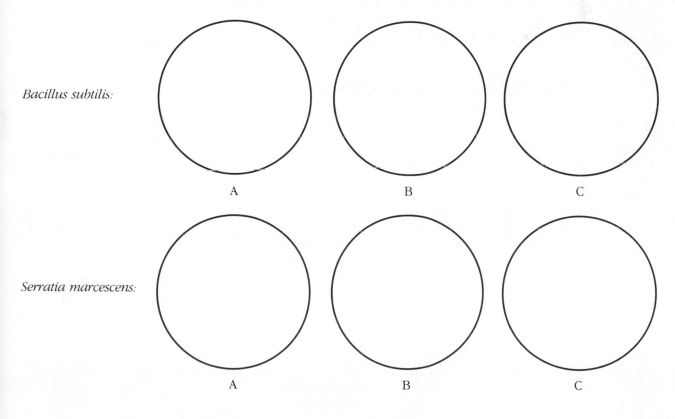

Bacillus subtilis:

A   B   C

Serratia marcescens:

A   B   C

Conclusion _____

_____

_____

_____

_____

Questions

1. If the *Bacillus* had sporulated before exposure to radiation, would that affect the results? _____

    _____

    _____

2. What are the two variables in ultraviolet radiation treatment? _____

    _____

3. Can dark repair be a factor in this experiment? _____

    _____

4. Ultraviolet radiation is used in hospital operating rooms; you've seen its limitations. What, then, is the object of

    using it? _____

    _____

    _____

    _____

5. Why are there still some colonies growing in the areas exposed to ultraviolet light? _____

    _____

    _____

Chemical Methods of Control: Disinfectants and Antiseptics

*One nineteenth-century method of avoiding
cholera: Wear a pouch of foul-smelling herbs
around your neck. If the odor is bad enough,
disease carriers will spare you the trouble of avoiding them.*

ANONYMOUS

Objectives

After completing this exercise you should be able to:

1. Define the following terms: disinfectant and antiseptic.
2. Evaluate the relative effectiveness of various chemical substances as antimicrobial agents.
3. Calculate a phenol coefficient.

Background

A wide variety of chemicals called **antimicrobial agents** are available for controlling the growth of microbes. Chemotherapeutic agents are used internally and will be evaluated in Exercise 25. **Disinfectants** are chemical agents used on inanimate objects to lower the level of microbes on that surface; **antiseptics** are chemicals used on living tissue to decrease the number of microbes. Disinfectants and antiseptics affect bacteria in many ways. Those that result in bacterial death are called **bactericidal agents.** Those causing temporary inhibition of growth are **bacteriostatic agents.**

No single chemical is the best to use in all situations. Antimicrobial agents must be matched to specific organisms and environmental conditions. Additional variables to consider in selecting an antimicrobial agent include pH, solubility, toxicity, organic material present, and cost. In evaluating the effectiveness of antimicrobial agents, the concentration, length of contact, and whether it is lethal *(-cidal)* or inhibiting *(-static)* are the important criteria. One method of measuring the effectiveness of a chemical agent is to determine its **phenol coefficient,** where the effectiveness of a phenolic-based chemical is compared to phenol. **Phenol** and its derivatives, called **phenolics,** kill bacteria by damaging the plasma membrane, inactivating enzymes, and denaturing proteins. The phenol coefficient is a com-

parative test using a standard organism such as *Staphylococcus aureus,* as shown in the sample problem (see the Sample Problem). The phenol coefficient is limited to bactericidal phenol-like compounds and cannot be used to evaluate bacteriostatic compounds.

In this exercise, we will perform a modified determination of a phenol coefficient.

Sample Problem: Phenol Coefficient

Purpose

To compare hexachlorophene to phenol and obtain the phenol coefficient of hexachlorophene.

Procedure

1. Serial dilutions of each chemical were set up: 1:10 to 1:500. How many tubes are needed? (See Appendix B.) _____
2. An equal amount of broth culture of a standard bacterium was added to each tube. Why is *Staphylococcus aureus* used to test hexachlorophene? _____
3. At 5-minute, 10-minute, and 15-minute intervals, tubes of sterile nutrient broth were inoculated with a loopful of each dilution.
4. The nutrient broth was incubated at 35°C for 24 to 48 hours and examined for growth.
5. Results were recorded as follows: (+) for growth and (−) for no growth; (−) means the agent has killed the bacteria.

Chemical	Dilution	Exposure Time (min)		
		5	10	15
Phenol	1:10–1:70	−	−	−
	1:80	+	−	−
	1:90	+	+	−
	1:100	+	+	−
	1:110–1:500	+	+	+
Hexachlo-rophene	1:10–1:150	−	−	−
	1:160	+	−	−
	1:170	+	+	+
	1:180	+	+	+
	1:190–1:500	+	+	+

Results

Interpretation

Circle the dilution of the test substance (hexachlorophene) that killed the bacteria in 10 minutes but not in 5 minutes. Circle the dilution of phenol that killed the bacteria in 10 minutes but not in 5 minutes.

Calculate the phenol coefficient as follows:

$$\frac{\text{Reciprocal of the hexachlorophene dilution circled}}{\text{Reciprocal of the phenol dilution circled}} = \text{Phenol coefficient}$$

Conclusions

1. What is the phenol coefficient of hexachlorophene? _____

2. In this sample, was hexachlorophene shown to be more or less effective than phenol? _____

3. How can you tell? _____

Materials

Glucose fermentation tubes (9)

Sterile water

Sterile tubes (3)

5-ml sterile pipettes (2)

1-ml sterile pipettes (2)

Phenol (0.5%)

Test substance: chemical agents such as bathroom cleaner, floor cleaner, mouthwash, lens cleaner, acne cream. Bring your own.

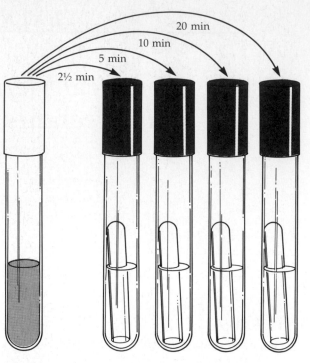

Chemical and bacteria   Glucose fermentation tubes

Figure 24.1 _____

Transfer a loopful from the tube containing phenol and *Staphylococcus* to a glucose fermentation tube at the time intervals shown. Repeat the procedure with a loopful from the tube containing the test substance and *Staphylococcus* into a different set of glucose fermentation tubes.

Culture

Staphylococcus aureus

Techniques Required

Exercises 10 and 14 and Appendix A.

Procedure

1. Using sterile water, prepare a dilution of the test substance in a sterile tube, diluted to the strength at which it is normally used. If a paste, it must be suspended in sterile water.

2. Transfer 5 ml of the diluted test substance to a sterile tube. Label the tube. Add 5 ml 0.5% phenol to a sterile tube. Label the tube.

3. Label four glucose fermentation tubes "Phenol" and four "Test." Then label one tube of each set "2.5"; label another pair "5"; continue with "10" and "20" (Figure 24.1). Label the remaining glucose fermentation tube "0."

4. Inoculate the 0 tube with a loopful of *S. aureus*.
5. Aseptically add 0.5 ml of the *S. aureus* culture to each tube prepared in step 2.
6. Transfer one loopful from each tube to a corresponding glucose fermentation tube at 2.5 minutes, 5 minutes, 10 minutes, and 20 minutes.
7. Incubate the glucose fermentation tubes at 35°C until the next lab period. (Discard the chemical/bacteria mixtures in the To Be Autoclaved basket.)
8. Observe the glucose fermentation tubes for growth and fermentation.

EXERCISE 24

Chemical Methods of Control: Disinfectants and Antiseptics

Name _____

Date _____

Lab Section _____

Purpose _____

_____

_____

Data

Time of exposure (min)	Control (Tube 0)			Chemical ___ Phenol ___			Chemical _____		
	Growth?	Acid?	Gas?	Growth?	Acid?	Gas?	Growth?	Acid?	Gas?
2.5									
5									
10									
20									

Conclusion _____

_____

_____

Questions

1. Was this a fair test? Is it representative of the effectiveness of the test substance? _____

_____

_____

2. How could the techniques be altered to measure bacteriostatic effects? _____

_____

3. Read the label of the preparation you tested. What is (are) the active ingredient(s)? _____

_____

Using your textbook or another reference, find the method of action of the test substance. _____

_____

_____

_____

4. What is the use-dilution method? _____

_____

_____

5. What does a phenol coefficient of 1 indicate? _____

_____

_____

Chemical Methods of Control: Antimicrobial Drugs

The aim of medicine is to prevent disease and prolong life; the ideal of medicine is to eliminate the need of a physician.

WILLIAM JAMES MAYO

Objectives

After completing this exercise you should be able to:

1. Define the following terms: antibiotic, chemotherapeutic agent, and MIC.
2. Perform an antibiotic sensitivity test.
3. Provide the rationale for the agar diffusion technique.

Background

The observation that some microbes inhibited the growth of others was made as early as 1874. Pasteur and others observed that infecting an animal with *Pseudomonas aeruginosa* protected the animal against *Bacillus anthracis.* Later investigators coined the word **antibiosis** (against life) for this inhibition and called the inhibiting substance an **antibiotic.** In 1928, Alexander Fleming observed antibiosis around a mold *(Penicillium)* growth on a culture of staphylococci. He found that culture filtrates of *Penicillium* inhibited the growth of many gram-positive cocci and *Neisseria* spp. In 1940, Selman A. Waksman isolated the antibiotic streptomycin, produced by an actinomycete. This antibiotic was effective against many bacteria that were not affected by penicillin. Actinomycetes remain an important source of antibiotics. Today, research investigators look for antibiotic-producing actinomycetes and fungi in soil, and have synthesized many antimicrobial substances in the laboratory. Antimicrobial chemicals absorbed or used internally, whether natural (antibiotics) or synthetic, are called **chemotherapeutic agents.**

A physician or dentist needs to select the correct chemotherapeutic agent intelligently and administer the appropriate dose in order to treat an infectious disease; then the practitioner must follow that treatment in order to be aware of resistant forms of the organism that might occur. The clinical laboratory isolates the **pathogen** (disease-causing organism) from a clinical sample and determines its sensitivity to chemotherapeutic agents.

In the **agar diffusion method,** bacteria to be tested are added to melted Mueller Hinton agar and poured onto solidified agar in a Petri plate. Paper disks impregnated with various chemotherapeutic agents are placed on the surface of the agar. During incubation, the chemotherapeutic agent *diffuses* from the disk, from an area of high concentration to an area of lower concentration. An effective agent will inhibit bacterial growth, and measurements can be made of the size of the **zones of inhibition** around the disks. The concentration of chemotherapeutic agent at the edge of the zone of inhibition represents its **minimum inhibitory concentration (MIC).** The MIC is determined by comparing the zone of inhibition with MIC values in a standard table (Table 25.1). The zone size is affected by such factors as the diffusion rate of the antibiotic and the growth rate of the organism. To minimize the variance between laboratories, the standardized **Kirby–Bauer test** for agar diffusion methods is performed in many clinical laboratories. This test uses Mueller Hinton agar. Mueller Hinton agar allows the chemotherapeutic agent to diffuse freely.

In this exercise, we will evaluate an agar-disk diffusion method.

Materials

Petri plate containing Mueller Hinton agar

Sterile cotton swab

Antibiotic dispenser and disks

Forceps

Alcohol

Garlic powder

Ruler

Table 25.1

Interpretation of Inhibition Zones of Test Cultures

Disk Symbol	Chemotherapeutic Agent	Diameter of Zones of Inhibition (mm)			
		Disk Content	Resistant	Intermediate	Susceptible
AM	Ampicillin[a] when testing gram-negative microorganisms and enterococci	10 μg	13 or less	14–16	17 or more
AM	Ampicillin[a] when testing staphylococci and penicillin G-susceptible microorganisms	10 μg	28 or less	—	29 or more
B	Bacitracin	10 units	8 or less	9–12	13 or more
CB	Carbenicillin when testing *Proteus* species and *Escherichia coli*	100 μg	19 or less	20–22	23 or more
CB	Carbenicillin when testing *Pseudomonas aeruginosa*	100 μg	13 or less	14–16	17 or more
CR	Cephalothin	30 μg	14 or less	15–17	18 or more
C	Chloramphenicol (Chloromycetic®)	30 μg	12 or less	13–17	18 or more
CC	Clindamycin[b] when reporting susceptibility to clindamycin	2 μg	14 or less	15–16	17 or more
CC	Clindamycin[b] when reporting susceptibility to lincomycin	2 μg	16 or less	17–20	21 or more
CL	Colistin[c] (Coly-mycin®)	10 μg	8 or less	9–10	11 or more
E	Erythromycin	15 μg	13 or less	14–17	18 or more
GM	Gentamicin	10 μg	12 or less	—	13 or more
K	Kanamycin	30 μg	13 or less	14–17	18 or more
ME	Methicillin[d]	5 μg	9 or less	10–13	14 or more
N	Neomycin	30 μg	12 or less	13–16	17 or more
NB	Novobiocin	30 μg	17 or less	18–21	22 or more
P	Penicillin G. when testing staphylococci[e]	10 units	20 or less	—	21 or more

Cultures (as assigned)

Staphylococcus aureus

Escherichia coli

Pseudomonas aeruginosa

Techniques Required

Exercise 10.

Procedure

1. Aseptically swab the assigned culture onto the appropriate plate. Swab in three directions to ensure complete plate coverage. Why is complete coverage essential? _____

_____

Let stand at least 5 minutes.

2. Follow procedure a or b.
   a. Place the chemotherapeutic impregnated disks by pushing the dispenser over the agar. Sterilize your loop and touch each disk to ensure better contact with the agar. Record the agents and the disk code in your Laboratory Report. Circle the corresponding chemicals in Table 25.1.
   b. Sterilize forceps by dipping in alcohol and burning off the alcohol.

Table 25.1 (continued)

Interpretation of Inhibition Zones of Test Cultures

Disk Symbol	Chemotherapeutic Agent	Diameter of Zones of Inhibition (mm)			
		Disk Content	Resistant	Intermediate	Susceptible
P	Penicillin G. when testing other microorganisms[e,f]	10 units	11 or less	12–21	22 or more
PB	Polymyxin B[c]	300 units	8 or less	9–11	12 or more
S	Streptomycin	10 μg	11 or less	12–14	15 or more
G	Sulfisoxazole	250 μg	12 or less	13–16	17 or more
ST	Sulfomethoxazole	250 μg	10 or less	11–15	16 or more
T	Tetracycline[g]	30 μg	14 or less	15–18	19 or more
VA	Vancomycin	30 μg	9 or less	10–11	12 or more

[a]The ampicillin disk is used for testing susceptibility of both ampicillin and betacillin.
[b]The clindamycin disk is used for testing susceptibility to both clindamycin and lincomycin.
[c]Colistin and polymyxin B diffuse poorly in agar and the accuracy of the diffusion method is less than with other antibiotics.
[d]The methicillin disk is used for testing susceptibility of all penicillinase-resistant penicillins.
[e]The penicillin G. disk is used for testing susceptibility to all penicillinase-susceptible penicillins except ampicillin and carbenicillin.
[f]This category includes some organisms such as enterococci and gram-negative bacilli that may cause systemic infections treatable with high doses of penicillin G.
[g]The tetracycline disk is used for testing susceptibility to all tetracyclines, e.g., chlorotetracycline, demeclocycline, doxycycline.

> ⚠ **Keep the beaker of alcohol away from the flame.**

Obtain a disk impregnated with a chemotherapeutic agent and place it on the surface of the agar (Figure 25.1*a*). Gently tap the disk to ensure better contact with the agar. Repeat, placing five to six disks the same distance apart on the Petri plate. See the location of the disks in Figure 25.1*b*. Record the agents and the disk code in your Laboratory Report. Circle the corresponding chemical in Table 25.1.

3. If you have space on your plate, add a small amount of garlic powder to an unused area.

4. Incubate inverted at 35°C until the next period. Measure the zones of inhibition in millimeters, using a ruler on the underside of the plate (Figure 25.1*b*). Record the zone size and, based on the values in Table 25.1, indicate whether the organism is sensitive, intermediate, or resistant. Observe the results of students using the other two bacteria. (See color plate VIII.1.)

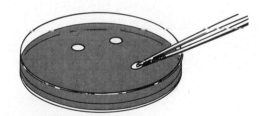

(a) Place disks impregnated with chemotherapeutic agents on an inoculated culture medium with sterile forceps to get the pattern shown in **(b).**

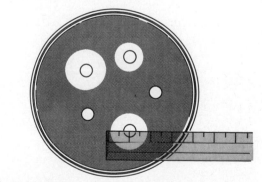

(b) After incubation, measure diameter of zone of inhibition.

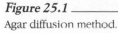

Figure 25.1

Agar diffusion method.

EXERCISE 25

Chemical Methods of Control: Antimicrobial Drugs

Name _____

Date _____

Lab Section _____

Purpose _____

_____

_____

Data

Effect of garlic powder? _____

Antimicrobial Agent	Disk Code	Staphylococcus aureus		Escherichia coli		Pseudomonas aeruginosa	
		Zone size	S, I, or R*	Zone size	S, I, or R*	Zone size	S, I, or R*
1.							
2.							
3.							
4.							
5.							
6.							
7.							
8.							

*S = sensitive; I = intermediate; R = resistant.

Questions

1. Which chemotherapeutic agent was most effective against each organism? _____

_____

2. What other factors are considered before using the chemotherapeutic agent in vivo? _____

_____

3. In which growth phase is an organism most sensitive to an antibiotic? _____

_____

4. Is the agar diffusion technique measuring bacteriostatic or bactericidal activity? _____

_____

_____

5. Why is the agar diffusion technique not a perfect indication of how the drug will perform in vivo? _____

_____

_____

6. What effect would the presence of tetracycline in the body have on penicillin therapy? _____

_____

_____

Effectiveness of Hand Scrubbing

*People are sick because they are poor, they become
poorer because they are sick and they become sicker
because they are poorer.*

ANONYMOUS

Objectives

After completing this exercise you should be able to:

1. Evaluate the effectiveness of hand washing and a surgical scrub.
2. Explain the importance of aseptic technique in the hospital environment.

Background

The skin is sterile during fetal development. After birth, a baby's skin is colonized by many bacteria for the rest of its life. As an individual ages and changes environments, the microbial population changes to match the environmental conditions. The microorganisms that are more or less permanent are called **normal flora.** Microbes that are present only for days or weeks are referred to as **transient flora.**

Discovery of the importance of hand and skin surface **disinfection** in disease prevention is credited to Ignatz Semmelweis at Vienna General Hospital in 1846. He noted that the lack of aseptic methods was directly related to the incidence of puerperal fever and other diseases. Medical students would go directly from the autopsy room to the patient's bedside and assist in child delivery without washing their hands. Less puerperal sepsis occurred in patients attended by nurses who did not touch cadavers. Semmelweis established a policy for the medical students of hand washing with a chloride of lime solution that resulted in a drop in the death rate due to puerperal sepsis from 12% to 1.2% in one year.

A layer of oil and the structure of the skin prevent the removal of all bacteria by hand washing. Soap helps remove the oil, and scrubbing will maximize the removal of bacteria. Hospital procedures require personnel to wash their hands before attending a patient, and a complete surgical scrub — removing the transient and many of the resident microflora — is done before surgery. Transient flora are usually removed after 10 to 15 minutes of scrubbing with soap. The surgeon's skin is never sterilized. Only burning or scraping it off would achieve that.

In this exercise, we will examine the effectiveness of washing skin with soap and water. Only organisms capable of growing aerobically on nutrient agar will be observed. Because organisms with different nutritional and environmental requirements will not grow, this procedure will involve only a minimum number of the skin microflora.

Materials

Petri plates containing nutrient agar (2)

Scrub brush

Bar soap or liquid soap (bring one from home)

Techniques Required

Exercise 9.

Procedure

1. Divide two nutrient agar plates into four quadrants. Label the sections of each plate 1 through 4. Label one plate "Water," the other "Soap." Which soap will you use? _____
2. Do the "Water" plate first. Touch section 1 with your fingers, wash well *without* soap, shake off excess water, and, while still wet, touch section 2. Do not dry your fingers with a towel. Wash again, and, while wet, touch section 3. Wash a final time, and touch section 4.
3. Use your other hand on the plate labeled "Soap." Wash with soap, rinse, shake off the excess water, then touch section 1.
4. Using a brush and soap, scrub your hand for 2 minutes, rinse, shake off the excess water, then touch section 2.

5. Using the brush, scrub your hand with soap for 5 minutes, rinse, shake off the excess water, then touch section 3. Repeat the soap and brush scrub for 10 minutes before touching section 4.

6. Incubate the plates inverted at 35°C until the next period.

7. Record the results.

EXERCISE 26

Effectiveness of Hand Scrubbing

Name _____

Date _____

Lab Section _____

Purpose _____
_____
_____

Data

Indicate the relative amounts of growth in each quadrant.

Section	Water Alone	Soap (type _____)
1. (No washing)		
2.		
3.		
4.		

Conclusion _____
_____
_____

Questions

1. What is a surgeon trying to accomplish with a 10-minute scrub with a brush followed by an antiseptic?

_____
_____
_____

2. How do normal flora and transient flora differ? _____
_____

3. If most of the normal flora and transient flora aren't harmful, then why must hands be scrubbed before

surgery? _____

_____

_____

4. Using your classmates' data, compare the results from bar soap and liquid soap. _____

_____

_____

PART 7

Microbial Genetics

All of the characteristics of bacteria—including growth patterns, metabolic activities, pathogenicity, and chemical composition—are inherited. These traits are transmitted from parent cell to offspring through genes. **Genetics** is the study of genes: how they carry information, how they are replicated and passed to the next generation or to another organism, and how their information is expressed within an organism to determine the particular characteristics of that organism.

The information stored in genes is called the **genotype.** All of the genes may not be expressed. **Phenotype** refers to the actual expressed characteristics, such as the ability to perform certain biochemical reactions (e.g., synthesis of a capsule or pigment).

Bacteria are invaluable to the study of genetics because large numbers can be cultured inexpensively, and their relatively simple genetic composition facilitates studying the structure and function of genes.

Bacteria undergo genetic change due to mutation or recombination. These genetic changes result in a change in the genotype. When a change occurs, in most instances we can expect that the product coded by certain genes will be changed. An enzyme coded by a changed gene may become inactive. This might be disadvantageous or even lethal if the cell loses a phenotypic trait it needs. Some genetic changes may be beneficial. For instance, if an altered enzyme coded by a changed gene has new enzymatic activity, the cell may be able to grow in new environments.

Genetic change in microbial populations is relatively easy to observe. In 1949, Howard B. Newcombe wrote:

> *Numerous bacterial variants are known which will grow in environments unfavorable to the parent strain, and to explain their occurrence two conflicting hypotheses have been advanced. The first assumes that the particular environment produces the observed change in some bacteria exposed to it, whereas the second assumes that the variants arose spontaneously during growth under normal conditions.**

Salvador Luria and Max Delbrück first demonstrated the latter **spontaneous mutation** hypothesis in 1943.

Genetic changes can be detected by using culturing methods that select for bacteria that have altered phenotypes. In Exercise 27, bacteria that are incapable of synthesizing essential amino acids will be selected. In this experiment, we will induce mutations by ultraviolet radiation.

Genetic recombination is the rearrangement of genes to form new combinations. Usually groups of genes from two different members of a species recombine or reshuffle, producing sequences containing nucleic acid from both participants and resulting in a new genetic variety (see the figure). In Exercise 28, we will isolate DNA from a bacterial culture. In Exercise 29, we will study genetic recombination resulting from transformation. In Exercise 30, we will investigate genetic recombination between bacterial genes and plant genes.

Cancer is thought to be the result of *permanent changes* in the nucleotide sequence of chromosomes. We will perform a test using principles of microbial genetics for identification of possible cancer-inducing substances in Exercise 31.

*Quoted in W. M. Stanley and E. G. Valens. *Viruses and the Nature of Life.* New York: Dutton, 1961, p. 8.

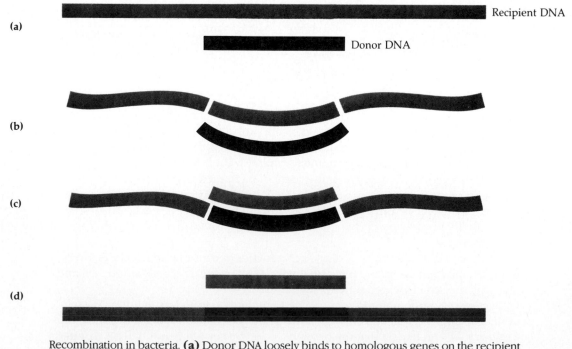

Recombination in bacteria. **(a)** Donor DNA loosely binds to homologous genes on the recipient DNA. **(b)** Recipient DNA is enzymatically cut. **(c)** Donor DNA replaces recipient DNA pieces. **(d)** DNA is rejoined.

Isolation of Bacterial Mutants

*Though coli may bother and vex us,
It's hard to believe they outsex us.
They accomplish seduction
by viral transduction
Foregoing the joys of amplexus.*

SEYMOUR GILBERT

Objectives

After completing this exercise you should be able to:

1. Define the following terms: prototroph, auxotroph, and mutation.
2. Differentiate between direct and indirect selection.
3. Isolate bacterial mutants by replica plating.

Background

For practical purposes, genes and the characteristics for which they code are stable. However, when millions of bacterial progeny are produced in just a few hours of incubation, genetic variants may occur. A variant is the result of a **mutation,** that is, a change in the sequence of nucleotide bases in the cell's DNA.

Metabolic mutants can be easily identified and isolated. Catabolic mutations could result in a bacterium that is deficient in the production of an enzyme needed to utilize a particular substrate. Anabolic mutations result in bacteria that are unable to synthesize an essential organic chemical. "Wild type," or nonmutated, bacteria are called **prototrophs,** and mutants are called **auxotrophs.** Auxotrophs have been isolated that either cannot catabolize certain organic substrates or cannot synthesize certain organic chemicals, such as amino acids, purine or pyrimidine bases, or sugars.

The culture used in this exercise is capable of synthesizing all of its growth requirements from glucose-minimal salts medium (Table 27.1). Mutations will be induced by exposing the cells to the mutagenic wavelengths of ultraviolet light (Exercise 23.) After exposure to ultraviolet radiation, auxotrophs that cannot grow on glucose-minimal salts medium will be identified.

Because of the low rate of appearance of recognizable mutants, special techniques have been developed to select for desired mutants. In the **gradient plate** (Exercise 21), **direct selection** is used to pick out mutant cells while rejecting the unmutated parent cells. We

Table 27.1
Composition of Glucose-Minimal Salts Agar

Glucose	0.1 g
Dipotassium phosphate	0.7 g
Monopotassium phosphate	0.2 g
Sodium citrate	0.05 g
Magnesium sulfate	0.01 g
Ammonium sulfate	0.15 g
Agar	1.5 g
Water	100 ml

will use **indirect selection** with the **replica-plating technique** in this exercise.

In replica plating, mutated bacteria are grown on a nutritionally complete solid medium. An imprint of the colonies is made on velveteen-covered or rubber-coated blocks and transferred to glucose-minimal salts solid medium and to a complete solid medium. Colonies that grow on the complete medium but not on the minimal medium are auxotrophs. Auxotrophs are identified *indirectly* because they will not grow.

Materials

Petri plates containing nutrient agar (complete medium) (4)

Petri plate containing glucose-minimal salts agar (minimal medium)

99-ml water dilution blanks (3)

Sterile 1-ml pipettes (4)

Ultraviolet lamp, 260 nm

Spreading rod ("hockey stick")

Sterile replica-plating block

Rubber band

Alcohol

Cultures (as assigned)

Serratia marcescens

Escherichia coli

Techniques Required

Exercises 10 and 23 and Appendices A and B.

Procedure*

First Period

1. Label the nutrient agar plates "A," "B," and "C" and the dilution blanks "1," "2," and "3."
2. Aseptically pipette 1 ml of the assigned broth culture to dilution blank 1 and mix well (Appendix A).
3. Using another pipette, transfer 1 ml from dilution blank 1 to dilution blank 2, as shown in Figure 27.1, and mix well. What dilution is in bottle 2? (Refer to Appendix B.) _____

4. Using another pipette, transfer 1 ml from dilution blank 2 to bottle 3 and 1 ml to the surface of plate A.
5. Mix bottle 3, and transfer 1 ml to the surface of plate B with a sterile pipette and 0.1 ml to the surface of plate C.
6. Disinfect a spreading rod ("hockey stick") by dipping in alcohol, quickly igniting the alcohol in a Bunsen burner flame, and letting the alcohol burn off.

> ⚠ **While it is burning, hold the hockey stick pointed *down*. Keep the beaker of alcohol away from the flame.**

Let the hockey stick cool.
7. Spread the liquid on the surface of plate C over the entire surface. Do the same with plates B and A (Figure 27.2).
8. Disinfect the hockey stick and return it. Why is it not necessary to disinfect the hockey stick between each plate when proceeding from plate C to B to A? _____

_____

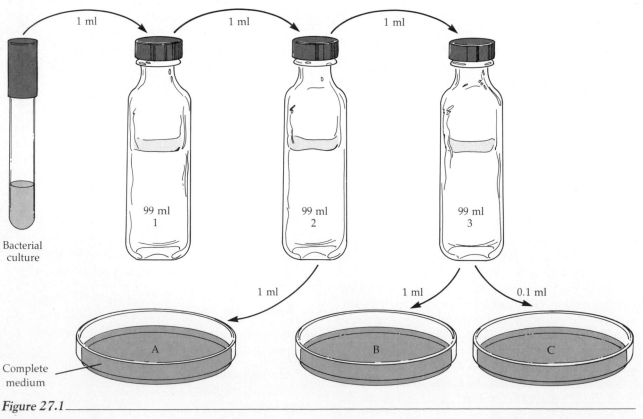

Figure 27.1 _____

Dilution procedure.

Figure 27.2 _____

Inoculation using a spreading rod ("hockey stick").

9.   Put the plate with the cover off, agar side up, about 5 cm from the ultraviolet lamp. Position the plate directly under the lamp. Turn the lamp on for 30 to 60 seconds, as assigned by your instructor.

⚠️ **Do not look directly at the light, and do not leave your hand exposed to it.**

---

10. Incubate the plates until the next period: *Escherichia coli* at 35°C; *Serratia marcescens* at 25°C.

Second Period

1.   Select a plate with 25 to 50 isolated colonies. Mark the bottom of the plate with a reference mark (Figure 27.3*a*). This is the master plate.

2.   Mark the uninoculated complete and minimal media with a reference mark on the bottom of each plate.

3.   If your replica-plating block is already assembled, proceed to step 4. If not, follow these instructions. Carefully open the package of velveteen. Place the replicator block on the center of the velveteen. Pick up the four corners of the cloth and secure tightly on the handle with a rubber band.

4.   Inoculate the sterile media by either step a or step b, as follows:

   a.   Hold the replica-plating block by resting the handle on the table with the rough surface, or velveteen, up (Figure 27.3*b*). Invert the master plate selected in step 1 on the block, and allow the master plate agar to lightly touch the block. Remove the cover from the minimal medium, align the reference marks, and touch the uninoculated minimal agar with the inoculated replica-plating block. Replace the cover. Remove the cover from the uninoculated complete medium and inoculate with the replica-plating block, keeping the reference marks the same.

   b.   Place the master plate and the uninoculated plates on the table. Place the reference marks in a 12 o'clock position and remove the covers. Touch the replica-plating block to the master plate, then, without altering its orientation, gently touch it to the minimal medium, then the complete medium (Figure 27.3*c*).

5.   Replace the covers, and incubate as before. Refrigerate the master plate.

6.   After incubation, compare the plates and record your results. (See color plate V.I.)

*Adapted from C. W. Brady. "Replica Plate Isolation of an Auxotroph." Unpublished paper. Whitewater, WI: University of Wisconsin, n.d.

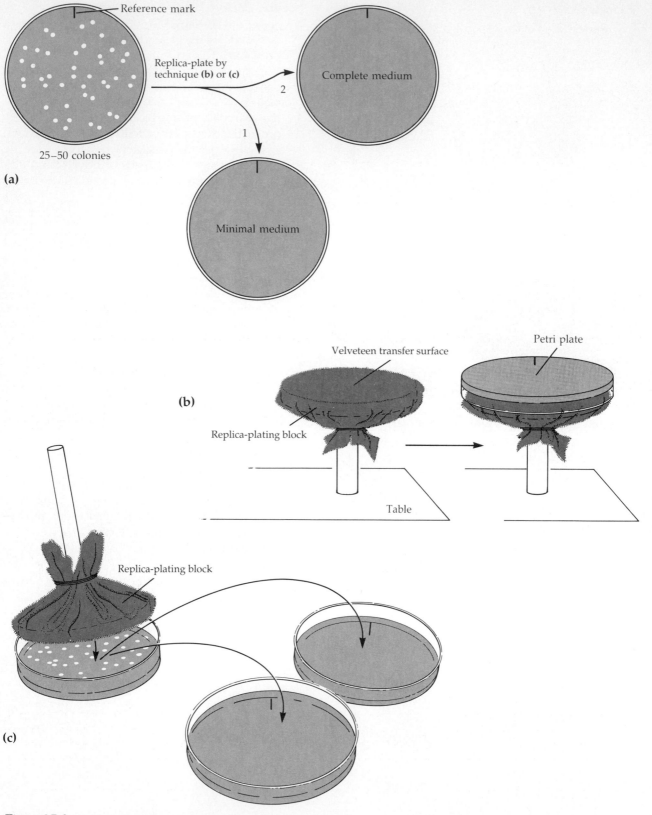

Figure 27.3

Replica plate. **(a)** Schematic diagram. **(b)** Transfer with
stationary replica-plating block. **(c)** Transfer by moving the
replica-plating block from plate to plate.

EXERCISE 27

Isolation of Bacterial Mutants

Name _____

Date _____

Lab Section _____

Purpose _____

_____

_____

Data

Mark the location of colonies and note any changes in pigmentation. Organism used: _____

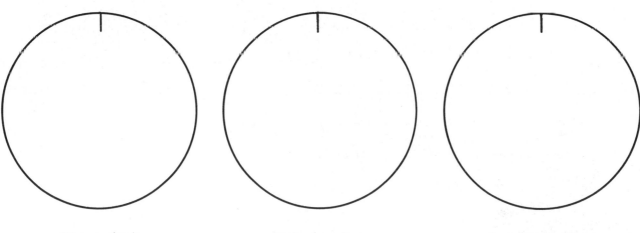

(Master plate)   Minimal medium
                        (Replica plated)   Complete medium
                                                  (Replica plated)

Class Data

S. marcescens	UV Exposure (sec)	Number of Auxotrophs	E. coli	UV Exposure (sec)	Number of Auxotrophs

Conclusions

1. What is the effect of longer exposure to ultraviolet light? _____

    _____

    _____

2. Compare the results for *E. coli* and *S. marcescens*. _____

    _____

    _____

Questions

1. Circle the auxotrophs on the diagram of the complete medium.

2. Why was the complete medium, as well as the minimal medium, replica plated? _____

    _____

    _____

3. How would you isolate a mutant from the replica plate? _____

    _____

    _____

    Does your technique involve a catabolic or anabolic mutant? _____

4. How would you identify the growth factor(s) needed by an auxotroph? _____

    _____

    _____

    _____

5. How would you use the replica-plating technique to isolate a mutant that is resistant to an antibiotic? _____

    _____

    _____

    _____

Isolation of DNA

Objectives

After completing this exercise you should be able to:

1. Explain the process of isolating DNA, then isolate DNA.
2. Determine the concentration of DNA in your extract.

Background

Scientists from many different disciplines have utilized the technique of genetic engineering to genetically alter bacteria and eucaryotic organisms. In **genetic engineering,** genes are isolated from one organism and incorporated by manipulation in a laboratory into other bacterial or eucaryotic cells.

The first step in genetic engineering is extraction of DNA from the cell that possesses the desired gene. To extract DNA from a bacterium, we disrupt the cell by chemical or mechanical means, or by osmotic pressure. The DNA is precipitated as a viscous, translucent mass with cold alcohol.

The DNA in the extract will be reacted with diphenylamine, resulting in a colored product. We can then determine the concentration of DNA in the extract by comparison with a standard curve, which is obtained by reacting known amounts of DNA with diphenylamine and plotting the absorbance of each concentration on a graph.

Materials

Centrifuge tube

5-ml pipettes (4)

Glass hook

95% ethyl alcohol, ice cold

Screw-capped tubes (7)

DNA standard (500 µg/ml)

Cuvette

Parafilm

Boiling chip

Ice

Boiling water bath

Diphenylamine

Centrifuge

Spectrophotometer

Culture

Halobacterium salinarum

Techniques Required

Appendices A and C.

Procedure

Isolation of DNA

1. Centrifuge the *Halobacterium* culture for 10 minutes at 3,000 rpm, to collect cells in the bottom of the tube.

> ⚠️ **Balance the centrifuge by placing a tube containing 10 ml water opposite your culture tube.**

> ☣️ **Discard the supernatant (liquid portion) by carefully pouring it into a container of disinfectant.**

2. Add 2 ml distilled water to the centrifuge tube and resuspend the pellet. Why did the appearance of the suspension change? _____
3. Gently add 4 ml cold ethyl alcohol by letting it run down the side of the tube. You should have two distinct layers. *Do not mix the alcohol and turbid aqueous layer.* Which layer is on top? _____
4. Collect DNA strands that appear at the interface by winding them gently onto a glass hook. (See color plate II.2.)
5. Resuspend the DNA in 2 ml distilled water in a screw-capped tube. Label the tube "7."

Standard Curve and Test for DNA

1. Start a boiling water bath.

> ⚠ **Be sure to add a boiling chip to the water. The boiling chip will prevent your test tubes from bumping and breaking.**

2. Label six screw-capped tubes, and add DNA stock solution and distilled water to each tube as shown below. The seventh tube contains your DNA.

Tube	Volume of Stock Solution	Distilled Water	Final [DNA] (μg/ml)
1	—	2.0 ml	0
2	0.4 ml	1.6 ml	100
3	0.8 ml	1.2 ml	200
4	1.2 ml	0.8 ml	300
5	1.6 ml	0.4 ml	400
6	2.0 ml	—	500

3. *Carefully* add 4.0 ml diphenylamine to each of the seven tubes. Cover each tube with Parafilm and carefully invert three times to mix.

> ⚠ **Diphenylamine contains concentrated sulfuric acid and concentrated acetic acid.**

4. Remove the Parafilm, place loose-fitting screw caps on the tubes, and then place the tubes in a beaker of vigorously boiling water for 10 minutes.
5. Transfer the tubes to an ice bath to cool.
6. Carefully transfer the contents of tube 1 to a cuvette and insert into a spectrophotometer. Adjust the absorbance to 0 (%T = 100) at 600 nm. (Refer to Appendix C.)
7. Transfer the contents of tube 2 to a cuvette and record the absorbance. Repeat with the remaining tubes (3 through 7).
8. Plot the standard curve for the DNA–diphenylamine reaction, using tubes 1 through 6.
9. Compare the absorbance of tube 7 to the standard curve to determine the concentration of DNA in your extract.
10. Discard all materials, as indicated by your instructor.

EXERCISE 28

Isolation of DNA

Name _____

Date _____

Lab Section _____

Purpose _____

_____

_____

Data

Describe the appearance, color, texture, and shape of your spooled DNA. _____

_____

_____

Tube	[DNA] µg/ml	Absorbance
1	0	
2	100	
3	200	
4	300	
5	400	
6	500	
7	Unknown	

Attach your DNA standard curve.

Conclusions

What is the concentration of DNA in your extract? _____

Questions

1. How were the *Halobacterium* cells disrupted in this exercise? _____

   _____

2. How would this procedure be changed to extract DNA from *E. coli*? _____

   _____

3. What physical characteristics of DNA allow it to be spooled into a glass rod? Why is it not possible to spool out

   precipitated proteins? _____

   _____

   _____

4. Why might you have a large spooled product and only a small amount of DNA? _____

   _____

5. Outline the procedures for locating a specific gene in this DNA. _____

   _____

   _____

   _____

   _____

Transformation and Genetic Engineering

Objectives

After completing this exercise you should be able to:

1. Define transformation, restriction enzyme, and ligation.
2. Analyze DNA by electrophoresis.
3. Accomplish genetic change through transformation.

Background

Transformation is a rare event involving acquisition by a *recipient* bacterium of small pieces of DNA released from a dead *donor* bacterium. The acquired pieces of DNA can give the recipient bacterium new characteristics. For genetic engineering of a bacterium using transformation, the desired gene is first inserted into a plasmid. Next, the recombinant plasmid is acquired by the bacterium. If the gene is expressed in the cell, the cell will produce a new protein product.

Genetic engineering is made possible by restriction enzymes. A **restriction enzyme** is an enzyme that recognizes and cuts only one particular sequence of nucleotide bases in DNA, and it cuts this sequence in the same way each time. We will use the restriction enzymes *Hin* dIII and *Bam* HI. *Hin* dIII, an enzyme isolated from *Haemophilus influenzae,* cuts double-stranded DNA at the arrows in the sequence

$$A \uparrow A G C T T$$
$$T T C G A \uparrow A$$

Bam HI, isolated from *Bacillus amyloliquefaciens,* cuts double-stranded DNA at the arrows in the sequence

$$G \uparrow G A T C C$$
$$C C T A G \uparrow G$$

In this exercise, we will use a plasmid containing an ampicillin-resistance gene (pAMP) and another plasmid with a kanamycin-resistance gene (pKAN) to transform an antibiotic-sensitive strain of *E. coli.* We will cut the two circular plasmids using the restriction enzymes (Figure 29.1). The DNA fragments will be ligated using DNA ligase, and the ligated DNA will be used to transform the *E. coli. E. coli* does not normally take up new DNA; its cell wall will be damaged by treatment with $CaCl_2$ so DNA can enter. We will use agarose gel electro-

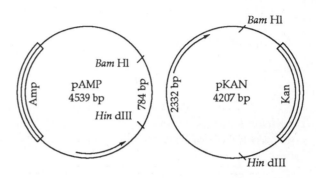

Figure 29.1 _____

Restriction maps of plasmids showing locations of antibiotic-resistance markers. DNA is measured in the number of base pairs (bp).

phoresis to determine whether the restriction enzymes cut the plasmids.

Materials

Plasmid Digestion

Sterile microcentrifuge tubes (2)

Micropipette, 2–10 μl

Sterile micropipette tips (4)

pAMP plasmid

pKAN plasmid

Restriction enzyme buffer

Bam HI and *Hin* dIII enzymes

Electrophoresis of Plasmids

Microcentrifuge tubes (3)

Micropipette, 2–10 μl

Sterile micropipette tips (7)

Casting tray and comb

Agarose, 0.8%

pAMP plasmid

Electrophoresis buffer

Tracking dye

Electrophoresis chamber and power supply

Ethidium bromide or methylene blue stain

Ligation of DNA Pieces

Sterile microcentrifuge tubes (2)

Micropipette, 2–10 μl

Sterile micropipette tips (5)

Ligation buffer

65°C water bath

Ice

Sterile distilled water

DNA ligase

Transformation of Competent amp^s/kan^s-E. coli Cells

Plate containing nutrient agar

Sterile microcentrifuge tubes (2)

Micropipettes, 2–10 μl and 200–1000 μl

Sterile micropipette tips (4)

50 mM CaCl$_2$

Ice

43°C water bath

Sterile nutrient broth

Culturing Transformed Cells

Micropipette, 10–100 μl

Sterile micropipette tips (2)

Plates containing nutrient agar (3)

Plates containing ampicillin-nutrient agar plates (3)

Plates containing kanamycin-nutrient agar plates (3)

Plates containing ampicillin and kanamycin-nutrient agar (2)

Hockey stick and alcohol

Culture

E. coli-amp^s/kan^s

Techniques Required

Exercises 10 and 27 and Appendices A and G.

Procedure (Figure 29.2)

A. Plasmid Digestion
Keep all reagents ice cold.

1. Label two sterile microcentrifuge tubes "1" and "2."
2. Aseptically add reagents as shown below.

Tube	pAMP	pKAN	Restriction Buffer	*Bam* HI/ *Hin* dIII
1	5.5 μl	—	7.5 μl	2 μl
2	—	5.5 μl	7.5 μl	2 μl

3. Close the caps and mix the tubes for 1 to 2 seconds by centrifuging.

> ⚠ **Be sure the centrifuge is balanced, by placing your tubes opposite each other.**

4. Incubate the tubes at 35°C for 20 minutes to 2 hours, as instructed. Tubes can be frozen until the next lab period.

B. Electrophoresis of Plasmids
1. Read about electrophoresis in Appendix G.
2. Use the melted agarose to pour a gel, as described in Appendix G.
3. Prepare three tubes as listed below. Remove samples aseptically from tubes 1 and 2. Save tubes 1 and 2 for procedure part C.

Tube	Uncut pAMP	Sample from part A tube #1	#2	Tracking dye
A	—	5 μl	—	1 μl
B	—	—	5 μl	1 μl
C	5 μl	—	—	1 μl

4. Mix tubes by centrifuging for 1 to 2 seconds.
5. Fill the electrophoresis chamber with electrophoresis buffer.
6. Load 5 μl from tube A into the first well of your gel. Load 5 μl from tube B into the second well and 5 μl from tube C into the third well.
7. Place the gel in the electrophoresis chamber, attach the lid, and apply 125V to the chamber. During the run, you will see the tracking dye migrate. The tracking dye migrates faster than the DNA. The dye is added so you can turn the power off before the DNA runs off the gel. Why do the two dyes separate? _____ Turn off the current when the two dyes have separated by 4 or 5 cm.

A. Plasmid Digestion

B. Electrophoresis of Plasmids

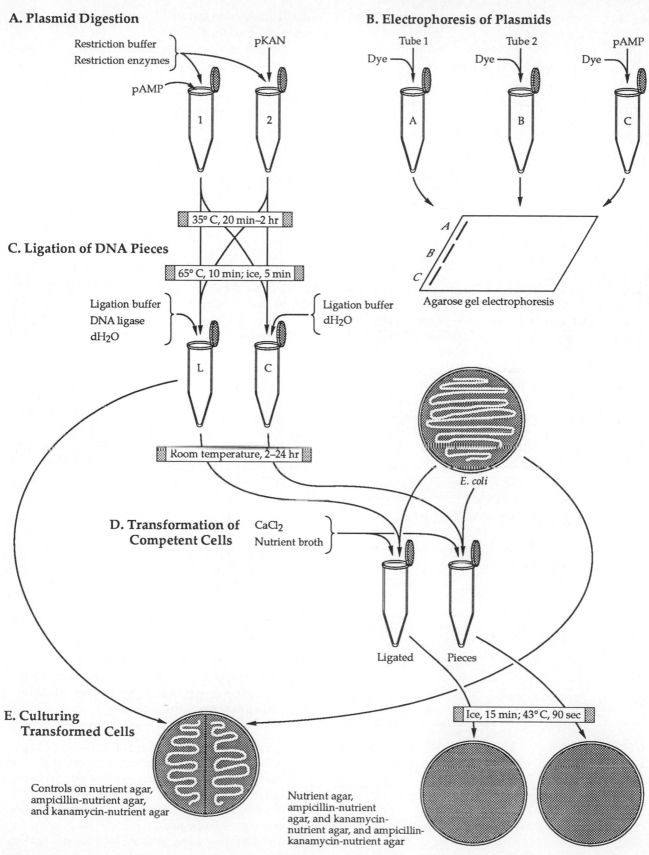

C. Ligation of DNA Pieces

D. Transformation of Competent Cells

E. Culturing Transformed Cells

Figure 29.2 _____

Procedure for plasmid digestion, electrophoresis, ligation, and transformation of *E. coli*.

8. To stain the gel, use either step a or step b below. (See color plate II.3.)
   a. Transfer the gel to the ethidium bromide staining tray for 5 to 10 minutes.

> ⚠️ **Do not touch the ethidium bromide. It is a mutagen.**

Transfer the gel to distilled water to destain for 5 minutes. (Chlorine in tap water will inactivate residual ethidium.) Place your gel on the transilluminator and close the plastic lid.

> ⚠️ **Do not look directly at the UV light. Do not turn the UV light on until the plastic lid is down.**

Ethidium bromide that is bound to DNA did not wash off and will fluoresce with UV light. Draw the bands or photograph your gel using the transilluminator camera.
   b. Transfer the gel to the methylene blue staining tray for 30 minutes to 2 hours. Destain by placing the gel in water for 30 minutes to overnight. Place your gel on a light box. Methylene blue that is bound to DNA did not wash off. Draw the bands.

C. Ligation of DNA Pieces

1. Label two sterile microcentrifuge tubes "L" for ligase and "C" for control.
2. Heat tubes 1 and 2 from part A at 65°C for 10 minutes to destroy the restriction enzymes. Why is this necessary? _____
3. Cool tubes 1 and 2 on ice for 5 minutes. Why is cooling necessary? _____
4. Aseptically add reagents to tubes L and C, as shown below.

Tube	Ligation Buffer	Sterile dH₂O	Tube 1	Tube 2	DNA Ligase
L	10 μl	3 μl	3 μl	3 μl	1 μl
C	10 μl	4 μl	3 μl	3 μl	—

5. Close caps securely and centrifuge for 1 to 2 seconds.
6. Incubate at room temperature for 2 to 24 hours.

D. Transformation of Competent ampˢ/kanˢ-E. coli Cells

1. Using a sterile loop, inoculate a nutrient agar plate with the ampˢ-kanˢ *E. coli*. Inoculate the plate with several loopfuls, streaking back and forth across the plate. Incubate the plate for 24 hours at 35°C.
2. Label two sterile microcentrifuge tubes "Ligated" and "Pieces." Aseptically pipette 250 μl of ice-cold CaCl₂ into each tube. Keep the tubes on ice.
3. With a sterile loop, transfer a loopful of *E. coli* to each tube. Mix to make a suspension.
4. Aseptically add 10 μl from tube L to the tube labelled "Ligated." Save tube L for part E. Add 10 μl from tube C to the "Pieces" tube.
5. Incubate the tubes on ice for 15 minutes to allow the DNA to settle on the cells.
6. *Heat shock* the cells by placing them in a 43°C water bath for 90 seconds so the DNA will enter the cells.
7. Return the tubes to the ice for 1 minute, then add 700 μl sterile nutrient broth. Incubate the tubes in a 35°C water bath for at least 3 hours.

E. Culturing Transformed Cells

1. Prepare the following controls. Inoculate one-half of a nutrient agar plate with a loopful of the ampˢ/kanˢ *E. coli* cells. Inoculate the other half with a loopful of the ligated plasmid DNA preparation (tube L). Similarly, inoculate an ampicillin-nutrient agar plate and a kanamycin-nutrient agar plate. How many plates do you have? _____
2. Add 100 μl from the "Ligated" tube to each of the following plates:

   Nutrient agar
   Ampicillin-nutrient agar
   Kanamycin-nutrient agar
   Ampicillin-kanamycin-nutrient agar

3. Sterilize a hockey stick by dipping in alcohol, quickly igniting the alcohol, and letting the alcohol burn off.

> ⚠️ **While it is burning, hold the hockey stick pointed down. Keep the beaker of alcohol away from the flame.**

Let the hockey stick cool.
4. Spread cells over one plate with the sterile hockey stick. Sterilize the hockey stick before spreading the cells over each of the remaining plates.
5. Repeat steps 2 through 4 to inoculate four plates from the "Pieces" tube.
6. Label the plates and incubate for 24 hours at 35°C.
7. Observe the plates and record your results.

EXERCISE 29

Transformation and Genetic Engineering

Name _____

Date _____

Lab Section _____

Purpose _____

_____

_____

Data

Sketch your electrophoresis results. Note the approximate sizes of the DNA fragments.

Controls	Growth on		
Inoculum	Nutrient Agar	Ampicillin-Nutrient Agar	Kanamycin-Nutrient Agar
amp^s/kan^s-*E. coli*			
DNA			

Transformation	Number of Colonies			
Inoculum	Nutrient Agar	Ampicillin-Nutrient Agar	Kanamycin-Nutrient Agar	Ampicillin-Kanamycin Nutrient Agar
Ligated				
Pieces				

Conclusions

1. Did you cut the plasmids? How do you know? _____

_____

2. What do the results of your transformation experiment indicate? _____

_____

_____

Questions

1. What is the purpose of each control? _____

_____

_____

_____

2. If there was no growth on the DNA/*E. coli* plates, what went wrong? _____

_____

_____

3. If you got growth on the DNA/*E. coli* plates, how could you rule out contamination? Mutation? _____

_____

_____

4. How could you prove ligation of the two plasmids occurred and not two separate transformation events? _____

_____

_____

_____

5. Design an experiment to determine whether *E. coli* DNA could be used to transform any other species. _____

_____

_____

_____

Agrobacterium: A Natural Genetic Engineer

Objectives

After completing this exercise you should be able to:

1. Describe the induction of crown gall by bacteria.
2. Describe how the Ti plasmid could be used in genetic engineering.

Background

Agrobacterium tumefaciens causes tumor growth in a wide variety of broad-leafed (dicotyledonous, or dicot) plants. The disease called **crown gall** is reportedly the third most serious plant disease in the United States.

Tumor-causing strains of *Agrobacterium* have a **tumor-inducing (Ti) plasmid.** A portion of the Ti plasmid, called **T-DNA,** becomes incorporated into the plant cell genome and is replicated along with the plant cell's chromosomes. T-DNA genes code for the production of unusual amino acid derivatives called *opines,* which the bacteria can metabolize, and auxin, a plant growth hormone that is necessary for the bacterium to bind to the plant cell. Moreover, the auxin causes rapid plant cell growth, resulting in a tumor. (See color plate II.1.) In short, *Agrobacterium* commandeers plant cells to produce chemicals for its growth.

Agrobacterium is found in soil. On contact with a host plant, the bacterium's cell wall lipopolysaccharide attaches to receptor sites on the cell walls of wounded plants. The bacterium does not enter the plant cell. Within the next few hours, the bacterium produces cellulose fibers to secure its attachment. Following infection, the T-DNA is integrated in the plant genome. Wounded dicots produce phenolics that induce expression of Ti genes.

Mutated Ti plasmids that do not induce tumors might be used as vectors to insert desirable genes into plants. The genes could be inserted into plant cells that, when cloned, produce plants that carry the new gene in their seeds.

In this exercise, we will observe the growth of *Agrobacterium* and the development of crown gall.

Materials

Petri plate containing nutrient agar

Carrot and/or tomato plants (2)

Sterile Petri plate with filter paper

Chlorine bleach

Isopropyl alcohol

Sterile water

Potato peeler

White tape or labels

Scalpel or razor blade

Mortar and pestle

Syringe containing sterile nutrient broth

Culture

Syringe containing *Agrobacterium tumefaciens*

Demonstrations

Slide of an *Agrobacterium*-induced gall

Large woody crown galls

Techniques Required

Exercises 5 and 11.

Procedure

1. Divide the nutrient agar plate in half. Place a drop of the *Agrobacterium* culture on the agar, and spread it over half of the agar with a sterile loop. Incubate inverted for 48 hours at room temperature, then refrigerate for use in step 3. Make a smear of the *Agrobacterium* and heat fix.

> **Do not recap the syringe.**

2. Inoculate either or both of the test plants as assigned by your instructor.
   a. Tomato: For the first plant, inject three or four nodes with about 0.05 ml *Agrobacterium*. Inject three or four nodes of the second plant with nutrient broth. Place a label around the petiole next to each site. Label and identify the

inoculation sites, as shown in Figure 30.1. Observe weekly. When good-sized galls appear, describe their appearance, and then cut them off.

b. Carrot: Wash the carrot well, and peel it to eliminate the outer surface; then dry, wash with chlorine bleach, and rinse with sterile water. Dip your scalpel in alcohol, burn off the alcohol, and cut four cross sections of carrot 5 to 8 mm thick. Place four slices on filter paper in a Petri plate, and wet with sterile water. Inoculate three of the slices with a drop of *Agrobacterium*. Incubate in your drawer, and examine and add water periodically. When galls appear, cut them off.

> ☣ **Discard syringes in the appropriate container.**

3. Slice a gall in half, making a longitudinal cut through the gall and plant tissue. Cut a thin slice of the cut edge of the gall, and transfer the slice intact to a microscope slide. Examine the gall with the low-power objective of your microscope, or use a dissecting microscope (Appendix E). Add a drop of safranin stain to the gall tissue to enhance the contrast.

4. Cut up the galls and grind them into a slurry in a mortar with sterile water. Inoculate the remaining half of the nutrient agar plate with the slurry. Incubate the plate as before. Make a smear from the slurry.

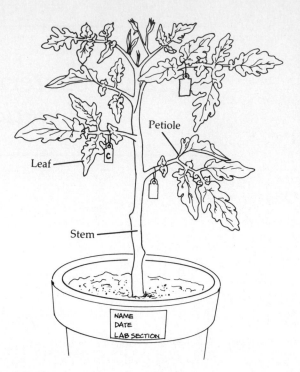

Figure 30.1 _____

Inoculate a tomato plant. Place labels close to the injection sites, as shown.

5. Gram stain both smears, and observe. Observe the growth on the plate and Gram stain.
6. Observe the demonstration galls and slide.

EXERCISE 30

Agrobacterium: A Natural Genetic Engineer

Name _____

Date _____

Lab Section _____

Purpose _____

_____

_____

Stains

Draw the appearance of your stains.

Gall
cross section

*Agrobacterium
tumefaciens*

Morphology: _____

Gram reaction: _____

Carrot or
tomato tumor

Morphology: _____

Gram reaction: _____

Demonstration

Morphology: _____

Gram reaction: _____

Cultures

Inoculum	Colony Description	Gram Stains
Agrobacterium		
Carrot		
Tomato		

Describe and sketch the galls at each laboratory period.

Questions

1. Did you find bacteria in the carrot or tomato gall? _____

2. What did this experiment indicate about the host range of *Agrobacterium?* _____

_____

3. Why aren't bacteria present in large quantities in galls? _____

_____

_____

4. How does *Agrobacterium* get transmitted to plants in nature? _____

5. Describe how a plant can be genetically engineered using *Agrobacterium*. _____

_____

_____

6. List some characteristics or genes that would be desirable to engineer into plants. _____

_____

_____

Ames Test for Detecting Possible Chemical Carcinogens

*The requirements of health can be stated
simply. Those fortunate enough to be born
free of significant congenital disease or
disability will remain well if three basic needs
are met: they must be adequately fed; they
must be protected from a wide range of
hazards in the environment; and they must
not depart radically from the pattern of personal
behavior under which man evolved,
for example, by smoking, overeating,
or sedentary living.*

THOMAS McKEOWN

Objectives

After completing this exercise you should be able to:

1. Differentiate between the terms mutagenic and carcinogenic.
2. Provide the rationale for the Ames test.
3. Perform the Ames test.

Background

Every day we are exposed to a variety of chemicals, some of which are **carcinogens;** that is, they can induce cancer. Historically, animal models have been used to evaluate the carcinogenic potential of a chemical, but the procedures are very costly and time-consuming, and result in the inadequate testing of some chemicals. Many chemical carcinogens induce cancer because they are **mutagens** that alter the nucleotide base sequence of DNA.

Bruce Ames and his co-workers at the University of California at Berkeley have developed a fast, inexpensive assay for mutagenesis using a *Salmonella* auxotroph. Most chemicals that have been shown to cause cancer in animals have proven mutagenic in the **Ames test.** Since not all mutagens induce cancer, the Ames test is a screening technique that can be used to identify high-risk compounds that must then be tested for carcinogenic potential. The Ames test utilizes auxotrophic strains of *Salmonella typhimurium,* which cannot synthesize the amino acid histidine (His⁻). The strains are also defective in *dark excision* repair of mu-

tations **(uvrB),** and an **rfa** mutation eliminates a portion of the lipopolysaccharide that coats the bacterial surface. The *rfa* mutation prevents the *Salmonella* from growing in the presence of sodium desoxycholate or crystal violet and increases the cell wall permeability; consequently, more mutagens enter the cell. The *uvrB* mutation minimizes repair of mutations; as a result, the bacteria are much more sensitive to mutations. To grow the auxotroph, histidine and biotin (due to the *uvrB*) must be added to the culture media.

In the Ames test, a small sample of the test chemical (a suspected mutagen) is either added to a melted soft agar (containing histidine) suspension of the *Salmonella* that is overlayed on minimal (lacking histidine) media or is placed in the center of minimal agar seeded with a lawn of the *Salmonella* auxotroph. (A small amount of histidine allows all the cells to go through a few divisions.) If certain mutations occur, the bacteria may *revert* to a wild type, or prototroph, and grow to form a colony. Only bacteria that have mutated (reverted) to His⁺ (able to synthesize histidine) will grow into colonies. In theory, the number of colonies is proportional to the mutagenicity of the chemical.

In nature many chemicals are neither carcinogenic nor mutagenic, but they are metabolically converted to mutagens by liver enzymes. The original Ames test could not detect the mutagenic potential of these in vivo (in the body) conversions. Suspect chemicals can be incubated with a suspension of rat liver enzyme preparation in an oxygenated environment to be "activated." *Salmonella* is exposed to the "activated" chem-

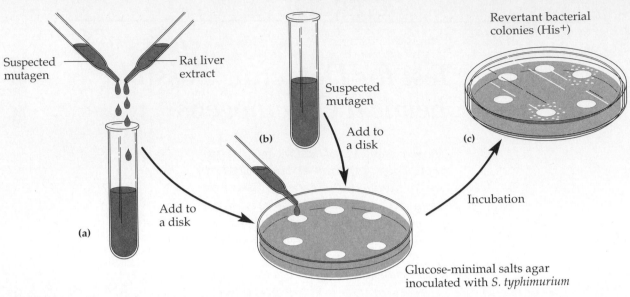

Figure 31.1 _____

The Ames test. **(a)** Rat liver extract can be used to activate a suspected mutagen. **(b)** A disk impregnated with a suspected mutagen is placed on the agar. The agar lacking histidine was inoculated with mutant *Salmonella* unable to synthesize histidine (His⁻). **(c)** Only bacteria that have further mutated (reverted) to His⁺ (able to synthesize histidine) will grow into colonies.

icals to test for mutagenicity (Figure 31.1*a*). In this exercise, the chemicals to be evaluated are mutagenic and will not require activation (Figure 31.1*b*).

> **⚠ *Salmonella typhimurium* is a potential pathogen.**

Materials

Petri plates containing glucose-minimal salts agar (Table 27.1) (2)

Tubes containing 2 ml soft agar (glucose-minimal salts + 0.05 mM histidine and 0.05 m*M* biotin) (2)

Sterile filter paper disks

Sterile 1-ml pipette

Forceps and alcohol

Paper disks soaked with suspected mutagens:
Benzo (α) pyrene
Ethidium bromide
Nitrous acid
2-aminofluorene
Cigarette smoke condensates
Food coloring
Saccharin
Hair dye
Household compound (your choice)

Culture

Salmonella typhimurium His⁻, *uvrB, rfa*

Techniques Required

Exercises 10 and 27.

Procedure*

1. Label one minimal-salts agar plate "Control" and the other "Test."
2. Aseptically pipette 0.1 ml *Salmonella* to one of the soft agar tubes, and quickly pour over the surface of the control plate. Tilt the plate back and forth gently to spread the agar evenly. Let harden in the dark. Why the dark? _____
3. Obtain another soft agar tube and repeat step 2 for the test plate. Why does the soft agar contain histidine? _____

> **⚠ Dip the forceps in alcohol, and, with the tip pointed down, burn off the alcohol. Keep the beaker of alcohol away from the flame.**

4. Aseptically place three to five disks saturated with the various suspected mutagens on the surface of the soft agar overlay of the test plate. If the chemical is crystalline, place a few crystals directly on the

*Adapted from D. J. Pringnitz. "The Effects of Ascorbic Acid on the Ability of Dimethylnitrosamine to Cause Reversion in a *Salmonella typhimurium* Mutant." Master's thesis. Mankato, MN: Mankato State University, 1975.

test plate. Label the bottom of the Petri plate with the name of each chemical.

> ⚠ *Be very careful.* **These are potentially dangerous compounds.**

5. Incubate both plates right-side up at 35°C for 48 hours.
6. Describe your results.

EXERCISE 31

Ames Test for Detecting Possible Chemical Carcinogens

Name _____

Date _____

Lab Section _____

Purpose _____

_____

_____

Data

Show the location of paper disks and bacterial growth on the plates. Number the disks to correspond to the table below.

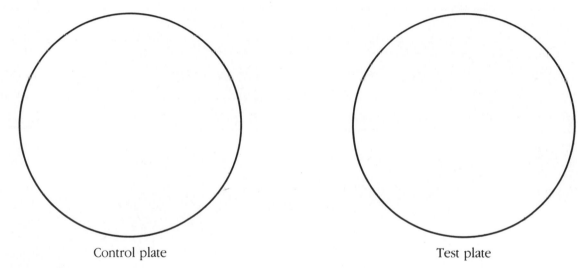

Control plate   Test plate

Complete the following table:

Compound	Growth? (large colonies)	Mutagenic?
1		
2		
3		
4		
5		

Questions

1. What degree of spontaneous reversion occurred? _____

_____

_____

2. Why are mutants used as test organisms in the Ames Test? _____

_____

_____

3. Does this technique give a minimum or maximum mutagenic potential? _____

_____

_____

4. What is the advantage of this test over animal tests? Disadvantages? _____

_____

_____

_____

PART 8

The Microbial World

In Part 8, we will examine some representative free-living microorganisms. In Exercise 32, we will identify an unknown bacterium using techniques and media described in Parts 1 through 5. In Exercises 33 through 36, we will examine eucaryotic microbes.

Organisms with eucaryotic cells include algae, protozoans, fungi, and higher plants and animals. The eucaryotic cell is typically larger and structurally more complex than the procaryotic cell. The DNA of a eucaryotic cell is enclosed within a membrane-bounded **nucleus.** In addition, eucaryotic cells contain membrane-bounded **organelles,** which have specialized structures and perform specific functions (see the figure).

The microorganisms studied in this part are all free-living chemoheterotrophs (Exercises 33, 34, and 36) except for the algae and cyanobacteria (Exercise 35), which are photoautotrophs.

Yeasts are possibly the best known microorganisms. They are widely used in commercial processes and can be purchased in the supermarket for baking. Yeasts are unicellular fungi (Exercise 33).

Leeuwenhoek was the first to observe the yeast responsible for fermentation in beer:

I have made divers observations of the yeast from which beer is made and I have generally seen that it is composed of globules floating in a clear medium (which I judged to be the beer itself).*

Many algae are only visible through the microscope, while others can be a few meters long. Algae are important producers of oxygen and food for protozoans and other organisms. A few unicellular algae, such as the agents of "red tides," *Gonyaulax catanella* and related species, are toxic to animals, including humans, when ingested in large numbers.

Protozoans, originally called "infusoria," were of interest to early investigators. In 1778, Friedrich von Gleichen studied food vacuoles by feeding red dye to his infusoria. A refinement of this experiment will be done in Exercise 36.

*Quoted in H. A. Lechevalier and M. Solotorovsky. *Three Centuries of Microbiology*. New York: Dover Publications, 1974, p. 502.

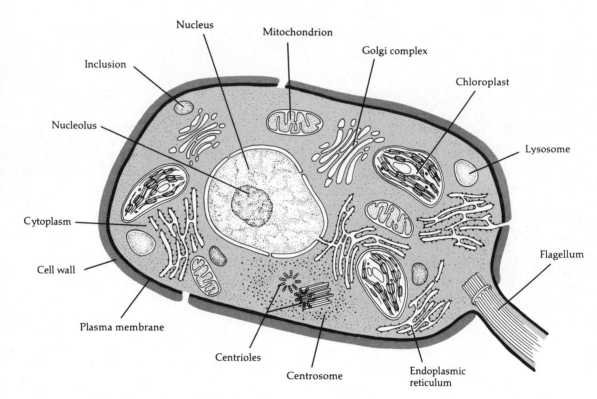

Highly schematic diagram of a composite eucaryotic cell.

EXERCISE 32

Unknown Identification

*The strategy of discovery lies in determining
the sequence of choice of problems to solve.
Now it is in fact very much more difficult to
see a problem than to find a solution to it.
The former requires imagination,
the latter only ingenuity.*

JOHN BERNAL

Objectives

After completing this exercise you should be able to:

1. Explain how bacteria are characterized and classified.
2. Use *Bergey's Manual*.
3. Identify an unknown bacterium.

Background

In microbiology, a system of classification must be available to allow the microbiologist to categorize and classify organisms. Communication among scientists would be very limited if no universal system of classification existed. The **taxonomy** (grouping) of bacteria is difficult because few definite anatomical or visual differences exist. With these limitations, most bacteria are characterized by evaluation of primary characteristics, such as morphology and growth patterns, and secondary characteristics, such as metabolism (Part 4) and serology (Part 11). A characteristic that is critical for distinguishing one bacterial group from another may be irrelevant for characterization of other bacteria.

The most important reference for bacterial taxonomy is *Bergey's Manual*,* in which bacteria are grouped into numbered **sections,** according to Gram stain reaction, cell shape, cell arrangements, oxygen requirements, motility, and nutritional and metabolic properties. Although the characteristics of a given group are relatively constant, through repeated laboratory culture, atypical bacteria will be found. This variability, however, only heightens the fun of classifying bacteria.

*J. G. Holt, ed. *Bergey's Manual of Systematic Bacteriology,* 1st ed. Baltimore: Williams & Wilkins, 1984–1989. See also *Bergey's Manual of Determinative Bacteriology,* 9th ed. Baltimore: Williams & Wilkins, 1992; and J. G. Holt, ed. *The Shorter Bergey's Manual of Determinative Bacteriology, 8th ed.* Baltimore: Williams & Wilkins, 1977.

You will be given an unknown heterotrophic bacterium to characterize and identify. By using careful deduction and systematically compiling and analyzing data, you should be able to identify the bacterium.

A key to species of bacteria is provided in Table 32.1. The key is an example of a dichotomous classification system; that is, a population is repeatedly divided into two parts until a description identifies a single member. Such a key is sometimes called an "artificial key" because there is no single correct way to write one. You may want to check your conclusion with the species description given in *Bergey's Manual.*

To begin your identification, ascertain the purity of the culture you have been given and prepare stock and working cultures. Avoid contamination of your unknown. Note growth characteristics and Gram stain appearance for clues about how to proceed. After culturing and staining the unknown, many bacterial groups can be eliminated. Final determination of your unknown will depend on careful selection of the relevant biochemical tests and weighing the value of one test over another in case of contradictions. (See color plate X.) Enjoy!

Materials

Petri plates containing trypticase soy agar (2)

Trypticase soy agar slant (2)

All stains, reagents, and media previously used

Culture

Unknown bacterium # _____

Techniques Required

Exercises 1, 2, 3–7, and 10–17.

Table 32.1
Key to the Identification of Selected Heterotrophic Bacteria

I. Cells are cocci

    A. Gram-positive

        1. Catalase produced

            a. Cells arranged in tetrads or glucose not fermented

                (1) Yellow pigment, hydrolyzes lipids — *Micrococcus luteus*

                (2) Red pigment, does not hydrolyze lipids — *M. roseus*

            b. Cells not in tetrads or glucose fermented

                (1) Acid from mannitol — *Staphylococcus aureus*

                (2) No acid from mannitol

                    (a) Acid from fructose — *S. epidermidis*

                    (b) No acid from fructose — *S. saprophyticus*

        2. Catalase not produced

            a. Growth in 6.5% NaCl or at pH 9.6

                (1) Acid from arabinose — *Enterococcus faecium*

                (2) No acid from arabinose — *E. faecalis*

            b. No growth in 6.5% NaCl or at pH 9.6

                (1) Acid from glycerol — *Streptococcus equisimilis*

                (2) No acid from glycerol

                    (a) Acid from inulin

                        aa. Acid from raffinose — *S. salivarius*

                        bb. No acid from raffinose — *S. sanguis*

                    (b) No acid from inulin — *S. mitis*

    B. Gram-negative

        1. Glucose fermented

            a. Nitrate reduced — *Neisseria mucosa*

            b. Nitrate not reduced — *N. sicca*

        2. Glucose not fermented — *N. flavescens*

II. Cells are rods

    A. Gram-positive

        1. Endospores present; catalase-positive

            a. Acid from mannitol

                (1) V–P positive — *Bacillus subtilis*

                (2) V–P negative — *B. megaterium*

            b. No acid from mannitol — *B. cereus*

(Continued)

Table 32.1
Key to the Identification of Selected Heterotrophic Bacteria

    2. No endospores; catalase-positive

        a. Acid from glucose; urease not produced   *Corynebacterium xerosis*

        b. No acid from glucose; urease produced   *C. pseudodiphtheriticum*

    3. No endospores; catalase-negative

        a. Acid and gas from glucose   *Lactobacillus fermentum*

        b. Acid, not gas, from glucose

            (1) Acid from mannitol   *L. casei*

            (2) No acid from mannitol   *L. acidophilus*

  B. Gram-negative

    1. Glucose not fermented; oxidase-positive

        a. Glucose oxidized

            (1) Litmus milk coagulated, peptonized, reduced   *Pseudomonas aeruginosa*

            (2) Litmus milk alkaline   *P. fluorescens*

        b. Glucose not oxidized

            (1) Gelatin hydrolyzed

                (a) White colonies   *Alcaligenes faecalis*

                (b) Yellow colonies   *A. paradoxus*

            (2) Gelatin not hydrolyzed   *A. aquamarinus*

    2. Glucose fermented; oxidase-negative

        a. Acid and gas from lactose

            (1) Citrate utilized

                (a) MR-negative, V–P positive   *Enterobacter aerogenes*

                (b) MR-positive, V–P negative   *Citrobacter freundii*

            (2) Citrate not utilized   *Escherichia coli*

        b. Lactose not fermented

            (1) Red pigment produced at 25°C   *Serratia marcescens*

            (2) No red pigment

                (a) Indole produced

                    aa. H_2S produced   *Proteus vulgaris*

                    bb. H_2S not produced   *Morganella morganii*

                (b) Indole not produced   *P. mirabilis*

Procedure

1. Streak your unknown onto the agar plates for isolation. Incubate one plate at 35°C and the other at room temperature for 24 to 48 hours. Note the growth characteristics and the temperature at which each one grew best.

2. Aseptically inoculate two trypticase soy agar slants. Incubate for 24 hours. Describe the resulting growth. Keep one slant culture in the refrigerator as your stock culture; the other is your working culture. Subculture onto another slant from your stock culture when your working culture is contaminated or not viable. Keep the working culture in the refrigerator when not in use.

3. Use your working culture for all identification procedures. When a new slant is made and its purity demonstrated, discard the old working culture. What should you do if you think your culture is contaminated? _____

4. Read the key in Table 32.1 to develop ideas on how to proceed. Perhaps determining staining characteristics might be a good place to start. What shape is it? _____
What can be eliminated? _____

5. After determining its staining and morphologic characteristics, determine which biochemical tests you will need. Do not be wasteful. Inoculate *only* what is needed. It is not necessary to repeat a test — do it once accurately. Do not perform unnecessary tests.

6. If you come across a new test — one not previously done in the course — ask if it is essential. Can you circumvent it? _____
If not, consult your instructor.

EXERCISE 32

Unknown Identification

Name _____

Date _____

Lab Section _____

Purpose _____

_____

Data

Write "not tested" next to tests that were not performed. Unknown # _____

Morphological, staining, and cultural characteristics	Sketches: Label and give magnification

The cell
- Shape _____
- Size _____
- Arrangement _____
- Endospores (position) _____
- _____
- Motility _____
-   Determined by _____
- Staining characteristics
- Gram _____ Age _____
- Other _____ Age _____

Colonies on Petri plate
- Diameter _____
- Appearance _____
- Color _____
- Elevation _____
- Margin _____
- Consistency _____

Agar slant   Age _____
- Amount of growth _____
- Form _____
- Consistency _____
- Color _____

Growth in broth   Age _____
- Amount of growth _____
- Pellicle _____
- Flocculant _____
- Sediment _____

Essential biochemical characteristics — Results

	Time (hr)						Temp. ___°C
	24	48	72	96	..	..	
Glucose							
Lactose							
Mannitol							
Catalase							
Oxidase							
H₂S							
Nitrate reduction							
Indole							
Methyl red							
V–P							
Citrate							

Abbreviations:
A = Acid
G = Gas
a = slight acid
alk = alkaline
+ = positive
− = negative
ng = no growth

Other characteristics, special media, etc. _____

Questions

1. What organism was in unknown # _____ ? _____

2. On a separate sheet of paper, write your rationale for arriving at your conclusion.

3. Why is it necessary to complete the identification of a bacterium based on its physiology rather than its

   morphology? _____

   _____

   _____

   _____

4. The following key lists bacterial groups as they correspond to sections in *Bergey's Manual.* Fill in the name of
   the Section indicated by the key.

    I. Gram-positive cell wall type   Section

        A. Endospore-forming   _____

        B. Nonendospore-forming

            1. Cocci   _____

            2. Rods   _____

                a. Acid-fast

                    (1) Mycelia formed   _____

                    (2) Mycelia not formed   _____

                b. Not acid-fast

                    (1) Mycelia formed   _____

                    (2) Mycelia not formed

                        (a) Regular   _____

                        (b) Club-shaped; stain irregularly   _____

    II. Gram-negative cell wall type

        A. Heterotrophic

            1. Helical or vibroid

                a. Possess axial filament   _____

                b. Possess flagella   _____

            2. Rods or cocci

                a. Aerobic   _____

                b. Facultatively anaerobic   _____

                c. Anaerobic

                    (1) Rods   _____

                    (2) Cocci   _____

    B. Autotrophic

        1. Photoautotrophic

            a.   Produce O_2 _____

            b.   Do not produce O_2 _____

        2. Chemoautotrophic

            a.   Oxidize N-containing compounds   _____

            b.   Oxidize S-containing compounds   _____

III.   No cell wall   _____

Fungi: Yeasts

Objectives

After completing this exercise you should be able to:

1. Culture and identify yeasts.
2. Differentiate between yeasts and bacteria.

Background

Fungi possess eucaryotic cells and can exist as unicellular or multicellular organisms. They are heterotrophic and obtain nutrients by absorbing dissolved organic material through their cell walls and plasma membranes. Fungi (with the exception of yeasts) are aerobic. Unicellular yeasts, multicellular molds (Exercise 34), and macroscopic species such as mushrooms are included in the Kingdom Fungi. Fungi generally prefer more acidic conditions and tolerate higher osmotic pressure and lower moisture than bacteria. They are larger than bacteria, with more cellular and morphologic detail. In contrast to bacterial characterization, primary characteristics, such as morphology and cellular detail, are used to classify fungi, with little attention given to secondary characteristics, such as metabolism and antigenic composition. Fungi are structurally more complex than bacteria but are less diverse metabolically.

Yeasts are nonfilamentous, unicellular fungi that are typically spherical or oval in shape. Yeasts are widely distributed in nature, frequently found on fruits and leaves as a white, powdery coating. Yeasts reproduce asexually by budding, a process in which a new cell forms as a protuberance (bud) from the parent cell (Figure 33.1a). In some instances, when buds fail to detach themselves, a short chain of cells called a **pseudohypha** forms (Figure 33.1b). When yeasts reproduce sexually, they may produce one of several types of sexual spores. The type of sexual spore produced by a species of yeast is used to classify the yeast as to division (phylum). (Divisions of fungi are described in Exercise 34.) Metabolic activities are also used to identify genera of yeasts.

Yeasts are facultative anaerobes (Exercise 19). Their metabolic activities are used in many industrial fermentation processes. Yeasts are used to prepare many foods, including bread, and beverages, such as wine and beer.

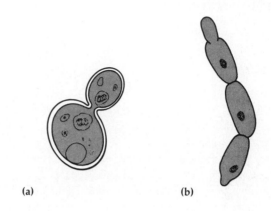

(a)   (b)

Figure 33.1 _____
Budding yeasts. **(a)** A bud forming from a parent cell.
(b) Pseudohyphae are short chains of cells formed by some yeasts.

In the laboratory, **Sabouraud agar,** a selective medium, is commonly used to isolate yeast. Sabouraud agar has very simple nutrients (glucose and peptone) and a low pH, which inhibits the growth of most other organisms. Many of the techniques useful in working with bacteria can be applied to yeasts.

Materials

Glucose fermentation tubes (2)

Sucrose fermentation tubes (2)

Petri plates containing Sabouraud agar (2)

Bottle containing glucose–yeast extract broth

Sterile cotton swab

Cover slip

Test tube

Methylene blue

Balloon

Fruit or leaves

Cultures (as assigned)

Baker's yeast

Rhodotorula rubra

Candida albicans

Saccharomyces cerevisiae

Techniques Required

Exercises 2, 11, 12, and 14.

Procedure

Yeasts

1. Gently suspend a pinch of baker's yeast in a small amount of lukewarm water in a test tube, creating a milky solution.
2. Each pair of students will use one of the yeast cultures and the suspension of baker's yeast.
   a. Divide one Sabouraud agar plate in half. Streak one half with a known yeast culture and the other half with the baker's yeast suspension.
   b. Inoculate each organism into a glucose fermentation tube and a sucrose fermentation tube.
3. Incubate all media at 35°C until growth is seen.

4. Make a wet mount (Exercise 2) of each culture by using a small drop of methylene blue. Record your observations.
5. After the yeast have grown, record your results. Examine cultures of the yeasts you did not culture, and record pertinent results. (See color plate IV.1.)

Yeast Isolation

1. Cut the fruit or leaves into small pieces. Place them in the bottle of glucose–yeast extract broth. Cover the mouth of the bottle with a balloon. Incubate the bottle at room temperature until growth has occurred. Record the appearance of the broth after incubation. Was gas produced? _____
2. Divide a Sabouraud agar plate in half. Each partner inoculates half of the medium, following either procedure a or procedure b.
   a. Swab the surface of your tongue with a sterile swab. Inoculate one-half of the agar surface with the swab. Why will few bacteria grow on this medium? _____
   b. Using a sterile inoculating loop, streak one-half of the agar surface with a loopful of broth from the bottle just prepared in step 1.
3. Incubate the plate inverted at room temperature until growth has occurred. Prepare wet mounts with methylene blue from different appearing colonies. Record your results.

EXERCISE 33

Fungi: Yeasts

Name _____

Date _____

Lab Section _____

Purpose _____

_____

Data

Record your data in the tables.

Yeasts

Fermentation tubes:

Organism	Glucose			Sucrose		
	Acid	Gas	Fermentation?	Acid	Gas	Fermentation?
Rhodotorula rubra						
Candida albicans						
Saccharomyces cerevisiae						
Baker's yeast						

Sabouraud agar plates:

Organism	Draw a Typical Colony	Wet Mounts
Rhodotorula rubra Color _____		
Candida albicans Color _____		

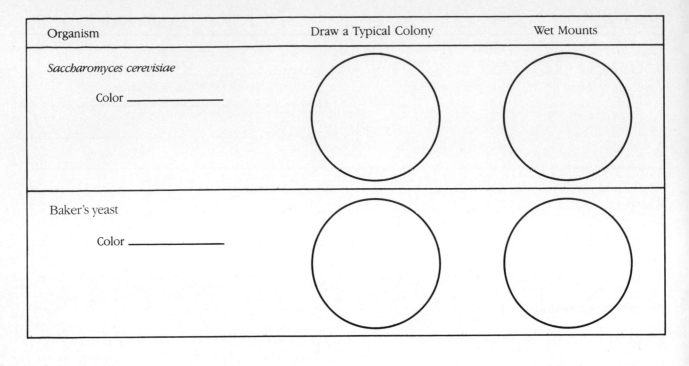

Organism	Draw a Typical Colony	Wet Mounts
Saccharomyces cerevisiae Color _____		
Baker's yeast Color _____		

Yeast Isolation

Plants used: _____

Describe the appearance of the glucose broth after _____ days' incubation. _____

_____

Was gas produced? _____

Sabouraud agar:

Colony appearance	Color	Size	Wet mounts
Mouth:			
Plants:			

Any bacteria seen? _____

If so, which colonies? _____

Questions

1. Could you identify the genus of baker's yeast? _____
_____

2. Did you culture yeast from your mouth? _____ From the plants? _____ How do you know? _____
_____

3. What was the purpose of the balloon on the glucose–yeast extract broth bottle? _____
_____

4. Define the term *yeast*. _____
_____
_____
_____
_____

5. Compare and contrast yeast and bacteria regarding their appearance both on solid media and under the
microscope. _____
_____
_____
_____

6. Name a genus of yeast that causes human disease. _____

7. Why are yeast colonies larger than bacterial colonies? _____
_____
_____
_____
_____

Fungi: Molds

*And what is weed? A plant whose virtues
have not been discovered.*

RALPH WALDO EMERSON

Objectives

After completing this exercise you should be able to:

1. Characterize and classify fungi.
2. Compare and contrast fungi and bacteria.
3. Identify common saprophytic molds.
4. Explain dimorphism.

Background

The multicellular filamentous fungi are called **molds.** Because of the wide diversity in mold morphology, morphology is very useful in classifying these fungi.

A macroscopic mold colony is called a **thallus** and is composed of a mass of strands called **mycelia.** Each strand is a **hypha,** with the **vegetative hyphae** growing in or on the surface of the growth medium. Aerial hyphae, called **reproductive hyphae,** originate from the vegetative hyphae and produce a variety of asexual reproductive **spores** (Figure 34.1). The hyphal strand of most molds is composed of individual cells separated by a crosswall, or **septum.** These hyphae are called **septate hyphae.** A few fungi, including *Rhizopus,* shown in Figure 34.1a, have hyphae that lack septa and are a continuous mass of cytoplasm with multiple nuclei. These are called **coenocytic hyphae.**

Fungi are characterized and classified by the appearance of their colony (color, size, and so on), hyphal organization (septate or coenocytic), and the structure and organization of reproductive spores. Because of the importance of colony appearance and organization, culture techniques and microscopic examination of fungi are very important.

Members of the division (phylum) **Zygomycota** are

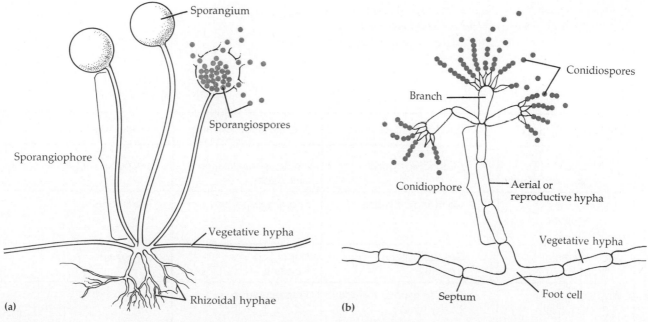

Figure 34.1

Asexual spores are produced by aerial hyphae. **(a)** Sporangiospores are formed within a sporangium. **(b)** Conidiospores are formed in chains. One possible arrangement is shown here.

saprophytic molds that have coenocytic hyphae. A **saprophyte** obtains its nutrients from dead organic matter; in healthy animals and plants, they do not cause disease. A common example is *Rhizopus* (bread mold). The asexual spores are formed inside a **sporangium,** or spore sac, and are called **sporangiospores** (Figure 34.1*a*). Sexual spores called **zygospores** are formed by the fusion of two cells.

The **Oomycota** are found in aquatic habitats and form sexual spores called **oospores,** formed by the fusion of two cells, and motile asexual spores called **zoospores.**

The **Ascomycota** include molds with septate hyphae and some yeasts. They are called sac fungi because their sexual spores, called **ascospores,** are produced in a sac, or **ascus.** The saprophytic molds usually produce **conidiospores** asexually (Figure 34.1*b*). The arrangements of the conidiospores are used to identify these fungi. Common examples are *Penicillium* and *Aspergillus.*

The **Basidiomycota** include the fleshy fungi, or mushrooms, and have sexual spores called **basiodiospores** produced by the club-shaped cap called a **basidium**.

A classification scheme for saprophytic fungi is shown in Table 34.1. A majority of the fungi classified in Table 34.1 are saprophytes, but each division contains a few genera that are **pathogens,** causing disease in plants and animals.

The **Deuteromycota** are also called the Fungi Imperfecti. These fungi are "imperfect" because sexual spores have not been demonstrated as yet. The Deuteromycota have septate hyphae and produce a variety of asexual spores. Most of the pathogenic fungi are, or once were, classified as Deuteromycota. When sexual spores are observed in members of a genus of Deuteromycota, the species is reclassified into a different genus (usually a new genus) and placed in the appropriate division.

Fungi, especially molds, are important clinically and industrially. Spores in the air are also the most common source of contamination in the laboratory. Sabouraud agar is a selective medium that is commonly used to isolate fungi (see Exercise 33).

Some pathogenic fungi exhibit **dimorphism;** that is, they have two growth forms. In pathogenic fungi, dimorphism is usually temperature dependent. The fungus is a yeast at 37°C and a mold at 25°C.

In this exercise, we will examine different molds and demonstrate dimorphism. *Mucor*'s dimorphism is not temperature dependent; it will be left for you to decide what physical condition affects *Mucor*'s growth.

Materials

Mold Culture
Petri plates containing Sabouraud agar (2)

Melted Sabouraud agar

Vaspar (one-half petroleum jelly and one-half paraffin)

Petri plate

Cover slips

Bent glass rod

Pasteur pipette

Dimorphic Gradient
Sabouraud agar, 5 ml, melted at 48°C

5-ml plastic microbeakers or small paper cups (2)

Razor blade

Tape

Table 34.1

Characteristics of Common Saprophytic Fungi

Division (Phylum)	Growth Characteristics	Asexual Reproduction	Sexual Reproduction
Oomycota "water molds"	Coenocytic hyphae	Motile sporangiospores called zoospores	Oospores
Zygomycota "conjugation fungi"	Coenocytic hyphae	Sporangiospores	Zygospores
Ascomycota "sac fungi"	Septate hyphae Yeastlike	Conidiospores Budding	Ascospores
Basidiomycota "club fungi"	Septate hyphae includes fleshy fungi (mushrooms)	Fragmentation	Basidiospores

Cultures (as assigned)

Rhizopus stolonifer

Aspergillis niger

Penicillium notatum

Mucor rouxii

Demonstrations

Prepared slides of the following:
  Zygospores
  Ascospores

Techniques Required

Exercises 2, 10, and 33 and Appendix E.

Procedure

First Period

1. Contaminate one Sabouraud agar plate in any manner you desire. Expose it to the air (outside, hall, lab, or wherever) for 15 to 30 minutes, or touch it. Incubate the plate inverted at room temperature for 5 to 7 days.

2. Inoculate the other Sabouraud agar plate with the mold culture assigned to you. Make one line in the center of the plate. Why don't you streak it? _____

   _____

   Incubate the plate inverted at room temperature for 5 to 7 days. (See color plates IV.2 and IV.3.)

3. Set up a slide culture of your assigned mold (Figure 34.2).
   a. Clean a slide and dry it.
   b. Place a drop of Sabouraud agar on the slide, flatten it to the size of a dime, and let it solidify.
   c. Using a sterile hot loop, carefully scrape off half of the circle of agar, leaving a smooth edge.
   d. Gently shake the mold culture to resuspend it, and inoculate the straight edge of the agar with a loopful of mold.
   e. Place a cover slip on the inoculated agar.
   f. Using a Pasteur pipette, add melted vaspar to seal three edges. Do not seal the inoculated edge.
   g. Put a piece of wet paper towel in the bottom of a Petri plate. Why? _____

      _____

   Place the slide culture on the glass rod on top of the towel and close the Petri plate. Incubate it at room temperature for 2 to 5 days. Do *not* invert the plate. What is the purpose of the glass rod? _____

   _____

   _____

(a) Place agar medium on the slide.

(b) Cut a straight edge on one side of the solidified agar.

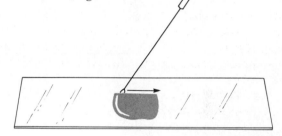

(c) Inoculate the mold onto the straight edge.

(d) Place a cover slip on the agar.

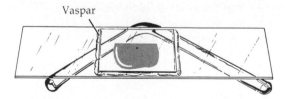

(e) Seal the three uninoculated edges with vaspar, and place the slide on a glass rod in a Petri plate.

Figure 34.2 _____
Preparation of a slide culture.

4. Observe the prepared slides showing sexual spores. Carefully diagram each of the spore formations in the space provided in the Laboratory Report.

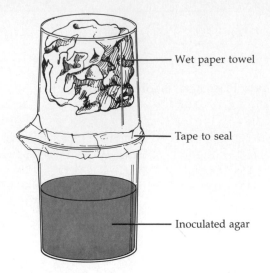

Wet paper towel

Tape to seal

Inoculated agar

Figure 34.3 _____
Tape the two beakers together as shown.

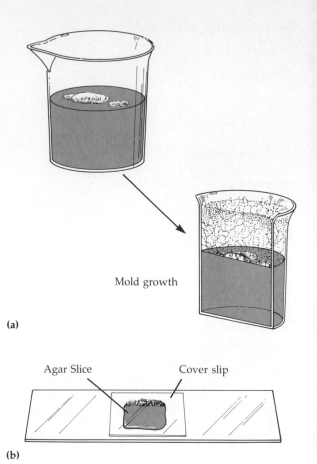

Mold growth

(a)

Agar Slice   Cover slip

(b)

Figure 34.4 _____
Observing fungal growth microscopically. **(a)** Cut the bottom beaker in half with a razor blade, and cut a vertical slice of agar. **(b)** Place the thin slice of agar on a slide, and cover with a cover slip.

Dimorphic Gradient*

1. Flick the fungal culture tube to resuspend the culture. Inoculate the melted agar with 2 or 3 loopfuls of the *Mucor* culture. Mix the tube by rolling it between your hands, and quickly pour the contents into an empty beaker before it hardens. Why doesn't the beaker need to be sterile? _____
_____

2. Place a piece of wet paper towel in the remaining empty beaker, and invert over the beaker containing the agar (Figure 34.3). Tape the edges.
3. Incubate at room temperature until growth occurs (5 to 7 days).

Second Period

1. Examine plate cultures of *each* mold (without a microscope), and describe their color and appearance. Then examine with a dissecting microscope (Appendix E). Look at the top and the underside.
2. Examine your contaminated Sabouraud plate and describe the results.
3. Examine slide cultures of each mold using a dissecting microscope. Record your observations.

*Adapted from S. Bertnicki-Garcia. "The Dimorphic Gradient of *Mucor rouxii:* A Laboratory Exercise." *ASM News* 38(9):486–488, 1972. Permission granted by the American Society for Microbiology.

4. To examine your dimorphic gradient, remove the tape and cut the beaker in half vertically with a razor blade (Figure 34.4a).
5. Cut a thin (approximately 1 mm) vertical slice of the inoculated agar with a razor blade. Carefully place the slice of agar on a slide and cover with a cover slip (Figure 34.4b).
6. Observe under low and high power. Scan the agar from the bottom of the slice to the top. Many focal planes are present.
7. Discard your beaker and disinfect your slide as instructed.

EXERCISE 34

Fungi: Molds

Name _____

Date _____

Lab Section _____

Purpose _____

_____

_____

Observations

Slide Cultures

Characteristics	Organism		
	Rhizopus stolonifer	*Aspergillus niger*	*Penicillium notatum*
Coenocytic?			
Type of spores			
Color of spores			
Diagram spore arrangement			

Sexual spores

Zygomycete   Ascomycete

Genus: _____   Genus: _____

_____ ✕   _____ ✕

Questions

1. Fill in the following table.

Organism	Asexual Spores	Sexual Spores	Division (Phylum)
Rhizopus			
Aspergillus			
Penicillium			

2. How do mold spores differ from bacterial endospores? _____

_____

_____

_____

3. What determines the form of fungi seen in the dimorphic gradient? _____

_____

_____

_____

4. Why do media used to culture fungi contain sugars? _____

_____

5. Why are antibiotics frequently added to Sabouraud agar for isolation of fungi from clinical samples? _____

_____

_____

6. How can fungi cause respiratory tract infections? _____

_____

_____

7. Why weren't pathogenic fungi used in this exercise? _____

_____

_____

Algae

Objectives

After completing this exercise you should be able to:

1. List criteria used to classify algae.
2. List general requirements for algal growth.
3. Compare and contrast algae with fungi and bacteria.

Background

Algae is the common name for photosynthetic organisms that lack true roots, stems, and leaves. Algae may be found in the ocean and in freshwater and on moist tree bark and soil. (See color plate IV.4.) Algae may be unicellular, colonial, filamentous, or multicellular. They exhibit a wide range of shapes: from the giant brown algae or kelp and delicate marine red algae to spherical green algal colonies. Algae are classified according to pigments, storage products, chemical composition of their cell walls, and flagella. The diversity of algae is illustrated by their classification: algal groups are found in two kingdoms.

Most freshwater algae belong to the groups listed in Table 35.1. Of primary interest to microbiologists are the **cyanobacteria.** Cyanobacteria have procaryotic cells and belong to the Kingdom Monera, along with other bacteria.

While algal growth is essential in providing oxygen and food for other organisms, some filamentous algae, such as *Spirogyra,* are a nuisance to humans because they clog filters in water systems (see color plate IV.5). Algae can be used to determine the quality of water. Polluted waters containing excessive nutrients from sewage or other sources have more cyanobacteria and fewer diatoms than clean waters. Additionally, the *number* of algal cells indicates water quality. More than 1,000 algal cells per milliliter indicates that excessive nutrients are present.

Materials

Pond water samples. (Distilled water may be added to the pond water periodically during incubation to replace water lost by evaporation.)

Table 35.1

Some Characteristics of Major Groups of Algae Found in Fresh Water

Characteristics	Algae classified as			
	Monera	Algal Protists		
	Cyanobacteria	Euglenoids	Diatoms	Green algae
Color	Blue-green	Green	Yellow-brown	Green
Cell wall	Bacteria-like	Lacking	Readily visible with regular markings	Visible
Cell type	Procaryote	Eucaryote	Eucaryote	Eucaryote
Flagella	Absent	Present	Absent	Present in some
Cell arrangement	Unicellular or filamentous	Unicellular	Unicellular or colonial	Unicellular, colonial, or filamentous
Nutrition	Autotrophic	Facultatively heterotrophic	Autotrophic	Autotrophic
Produce O_2	Yes; some use bacterial photosynthesis	Yes	Yes	Yes

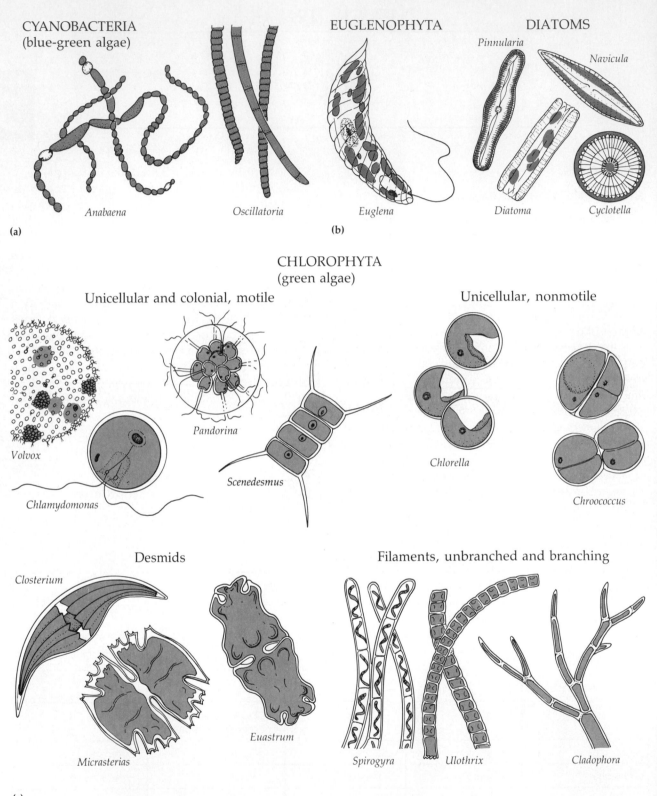

Figure 35.1 _____

Representative freshwater algae. **(a)** Representatives of the Kingdom Monera (Procaryota) found in fresh water. **(b)** and **(c)** Some members of the Kingdom Protista found in fresh water.

A. Incubated in the light for 4 weeks.
B. Incubated in the dark for 4 weeks.
C. Nitrates and phosphates added; incubated in the light for 4 weeks.
D. Copper sulfate added; incubated in the light for 4 weeks.

Techniques Required

Exercise 2.

Procedure

1. Prepare a hanging-drop slide from a sample of pond water A. Take your drop from the bottom of the container. Why? _____
_____

2. Examine the slide using the low and high-dry objectives. Identify the algae present in the pond water. Use Figure 35.1 for identification. Draw those algae that you cannot identify. Record the relative amounts of each type of alga from 4+ (most abundant) to + (one representative seen).

3. Repeat the observation and data collection for the remaining pond water samples.

EXERCISE 35

Algae

Name _____

Date _____

Lab Section _____

Purpose _____

_____

_____

Data

Name of Alga	Relative Abundance in Pond Water			
	Incubated in Light	Incubated in Dark	Incubated with NO_3^- and PO_4^{3-}	Incubated with $CuSO_4$
Drawings of other algae seen				

As a class, tabulate the number and various groups of algae observed in the following situations:

Light vs. dark

Nitrate vs. no nitrate (sample A)

Copper sulfate vs. no copper sulfate (sample B)

Conclusions

1. What can you conclude regarding the effects of light on algal growth? _____

_____

2. What can you conclude regarding the effects of the addition of nitrates and phosphates to pond water? _____

_____

_____

3. What can you conclude regarding the effects of the addition of copper sulfate to pond water? _____

_____

_____

Questions

1. Why can algae be considered indicators of productivity as well as of pollution? _____

_____

_____

2. How can algae be responsible for the production of more oxygen than land plants? _____

_____

_____

3. Why aren't algae included in the Kingdom Plantae? _____

_____

_____

_____

4. Describe one way in which algae and fungi differ. _____

_____

_____

How are they similar? _____

_____

_____

_____

EXERCISE 36

Protozoans

Objectives

After completing this exercise you should be able to:

1. List three characteristics of protozoans.
2. Explain how protozoans are classified.

Background

Protozoans are unicellular eucaryotic organisms that belong to the kingdom Protista. They are found in areas with a large water supply. Many protozoans live in soil and water, and some are normal flora in animals. A few species of protozoans are parasites.

Protozoans are aerobic heterotrophs. They feed on other microorganisms and on small particulate matter. Protozoans lack cell walls; in some (flagellates and ciliates), the outer covering is a thick, elastic membrane called a **pellicle.** Cells with a pellicle require specialized structures to take in food. The **contractile vacuole** may be visible in some specimens. This organelle fills with fresh water and then contracts to eliminate excess fresh water from the cell, allowing the organism to live in low solute environments. Would you expect more contractile vacuoles in freshwater or marine protozoans? _____

Protozoans may be classified on the basis of their method of motility. In this exercise, we will examine free-living members of three phyla of protists. The **Sarcodina,** or **amoebas** (Figure 36.1*a*), move by extending lobelike projections of cytoplasm called **pseudopods.** As pseudopods flow from one end of the cell, the rest of the cell flows toward the pseudopods.

The **Mastigophora,** or **flagellates** (Figure 36.1*b*), have one or more flagella. Although flagellates are heterotrophs, the organism used in this exercise is a facultative heterotroph. It grows photosynthetically in the presence of light and heterotrophically in the dark.

The **Ciliata** (Figure 36.1*c*) have many cilia extending from the cell. In some ciliates, the cilia occur in rows over the entire surface of the cell. In ciliates that live attached to solid surfaces, the cilia occur only around the oral groove. Why only around the oral groove? _____

Food is taken into the **oral groove** through the cytostome (mouth) and into the cytopharynx, where a **food vacuole** forms.

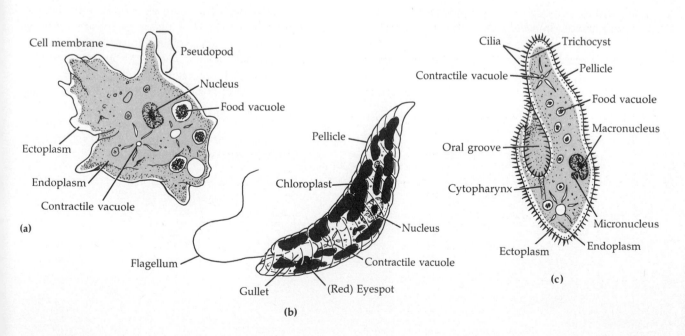

Figure 36.1 _____

Protozoans. **(a)** *Amoeba* moves by extending pseudopods. **(b)** *Euglena* has a whiplike flagellum. **(c)** *Paramecium* has cilia over its surface.

Materials

Methylcellulose, 1.5%

Acetic acid, 5%

Pasteur pipettes

Toothpicks

Cultures

Amoeba

Paramecium

Paramecium feeding on Congo red–yeast suspension

Euglena

Techniques Required

Exercise 2.

Procedure

1. Prepare a wet mount of *Amoeba*. Place a drop from the bottom of the *Amoeba* culture on a slide. Place one edge of the cover slip into the drop and let the fluid run along the cover slip (Figure 36.2). Gently lay the cover slip over the drop. Observe the amoeboid movement, and diagram it. Which region of the cytoplasm has more granules: the ectoplasm or the endoplasm? (Refer to Figure 36.1*a*).

2. Prepare a wet mount of *Euglena*. Follow one individual and diagram its movement. Can you see *Euglena's* red "eyespot"? _____
   Why do you suppose it is present in photosynthetic strains and not in nonphotosynthetic strains? ____
   _____
   Allow a drop of acetic acid to seep under the cover slip (Figure 36.3). How does *Euglena* respond?
   _____

3. Prepare a wet mount of *Paramecium* and observe its movement. (See color plate IV.5.) Why do you suppose it rolls and *Amoeba* does not? _____
   _____

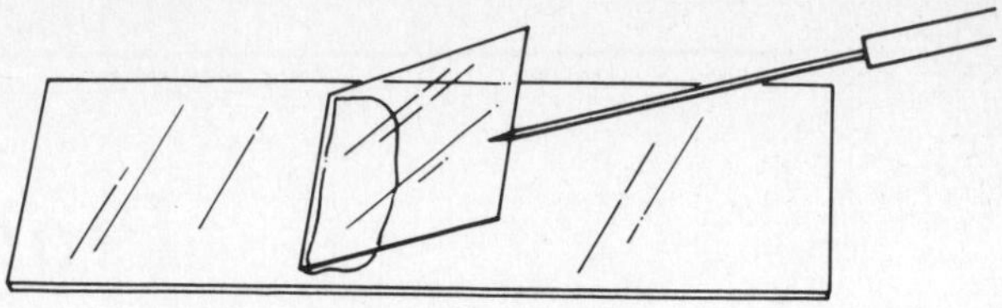

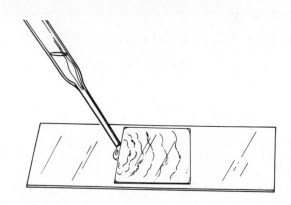

Figure 36.2 _____
Gently lower the cover slip.

Figure 36.3 _____
Add a drop of acetic acid at one edge of the cover slip. Allow it to diffuse into the wet mount.

4. Place a drop of methylcellulose on a slide. Make a wet mount of the *Paramecium* culture that has been feeding on a Congo red–yeast suspension in this mixture. The *Paramecium* will move more slowly in the viscous methylcellulose. Observe the ingestion of the red-stained yeast cells by *Paramecium*. Congo red is a pH indicator. As the contents of food vacuoles are digested, the indicator will turn blue (pH 3). What metabolic products would produce acidic conditions in the vacuoles? _____
   _____
   Count the number of red and blue food vacuoles in a *Paramecium*. Sketch a *Paramecium* and identify the locations of the food vacuoles.

EXERCISE 36

Protozoans

Name _____

Date _____

Lab Section _____

Purpose _____

_____

_____

Observations

Use a series of diagrams to illustrate the following:

The movements of *Amoeba* across the field of vision. _____ ×

The movement of *Euglena* and its flagellum. _____ ×

The ingestion of food, formation of a food vacuole, and movement of food vacuoles in *Paramecium*. Note any color changes in the food vacuoles. _____ ×

Questions

1. How did *Euglena* respond to the acetic acid? _____

_____

2. Which organism observed in this exercise would you bet on in a race? _____

3. Describe the arrangement of cilia on *Paramecium*. _____

_____

4. What criteria are used to place an organism in the Kingdom Protista? _____

_____

5. How are protozoans classified as to phyla? _____

_____

6. Why is *Euglena* often used to study algae (Exercise 35) and protozoans (Exercise 36)? _____

_____

_____

PART 9

Viruses

Viruses are fundamentally different from the bacteria studied in previous exercises, as well as from all other organisms. Wendell M. Stanley, who was awarded the Nobel Prize in 1946 for his work on tobacco mosaic virus, described the virus as "one of the great riddles of biology. We do not know whether it is alive or dead, because it seems to occupy a place midway between the inert chemical molecule and the living organism."*

Viruses are submicroscopic, filterable, infectious agents (see the figure). Viruses are too small to be seen with a light microscope and can be seen only with an electron microscope. Filtration is frequently used to separate viruses from other microorganisms. A suspension containing viruses and bacteria is filtered through a membrane filter with a small pore size (0.45 μm) that retains the bacteria but that allows viruses to pass through.

Viruses are *obligate intracellular parasites* with a simple structure and organization. Viruses contain DNA or RNA enclosed in a protein coat. Viruses exhibit quite strict specificity for host cells. Once inside the host cell, the virus may use the host cell's synthetic machinery and material to cause synthesis of virus particles that can infect new cells.

*Quoted in H. A. Lechevalier and M. Solotorovsky. *Three Centuries of Microbiology.* New York: Dover Publications, 1974, p. 25.

Viruses are widely distributed in nature and have been isolated from virtually every eucaryotic and procaryotic organism. Based on their host specificity, viruses are divided into general categories, such as **bacterial viruses (bacteriophages), plant viruses,** and **animal viruses.** A virus's host cells must be grown in order to culture viruses in a laboratory.

In Exercise 37, we will isolate bacteriophages and estimate the number of phage particles. In Exercise 38, we will isolate plant viruses on plant leaves.

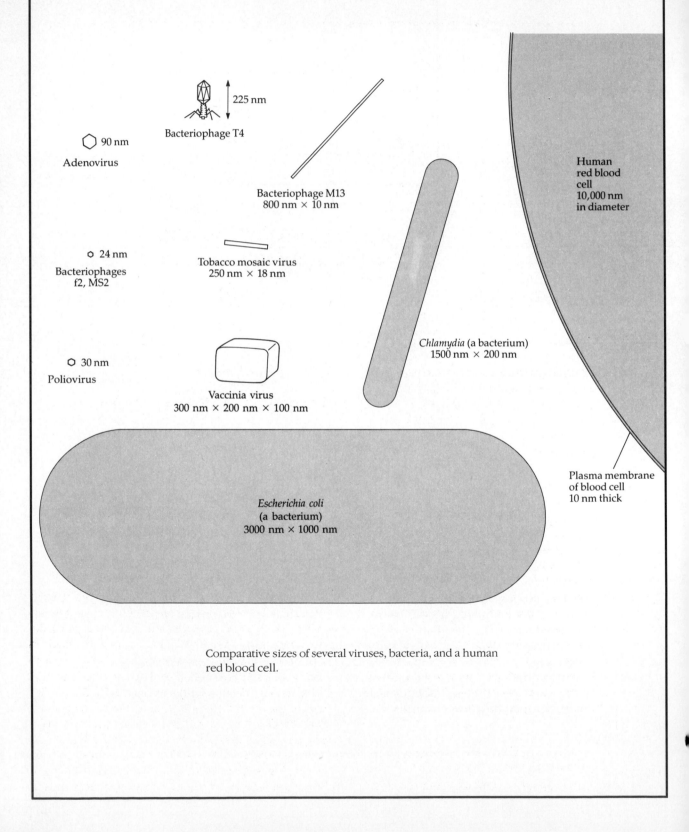

Comparative sizes of several viruses, bacteria, and a human red blood cell.

Isolation and Titration of Bacteriophages

Objectives

After completing this exercise you should be able to:

1. Isolate a bacteriophage from a natural environment.
2. Describe the cultivation of bacteriophages.
3. List the steps in the multiplication of a bacteriophage.
4. Determine the titer of a bacteriophage sample using the broth-clearing and plaque-forming methods.

Background

Bacteriophages, a term coined around 1917 by Félix d'Hérelle meaning "bacteria eater," parasitize most, if not all, bacteria in a very specific manner. Some bacteriophages, or **phages,** such as T-even bacteriophages, have a **complex structure** (Figure 37.1*a*). The protein coat consists of a polyhedral head and a helical tail, to which other structures are attached. The head contains the nucleic acid. To initiate an infection, the bacteriophage **adsorbs** onto the surface of a bacterial cell by means of its tail fibers and base plate. The bacteriophage injects its DNA into the bacterium during **penetration** (Figure 37.1*b*). The tail sheath contracts, driving the core through the cell wall and injecting the DNA into the bacterium.

Bacteriophages can be grown in liquid or solid cultures of bacteria. The use of solid media makes location of the bacteriophage possible by the **plaque-forming method.** Host bacteria and bacteriophages are mixed together in melted agar, which is then poured into a Petri plate containing hardened nutrient agar. Each bacteriophage that infects a bacterium multiplies, and releases several hundred new viruses. The new viruses infect other bacteria, and more new viruses are produced. All the bacteria in the area surrounding the original virus are destroyed, leaving a clear area, or **plaque,** against a confluent "lawn" of bacteria. The lawn of bacteria is produced by the growth of uninfected bacterial cells.

In this exercise, we will isolate a bacteriophage from host cells in a natural environment (i.e., in sewage and houseflies). Since the numbers of phages in a natural source are low, the desired host bacteria and additional

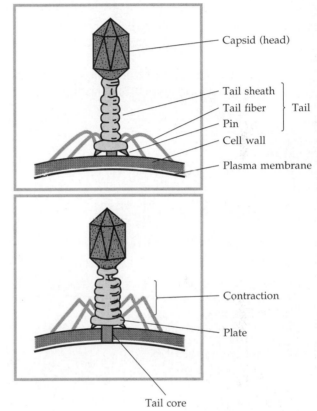

Figure 37.1

T-even bacteriophage. **(a)** Diagram of a bacteriophage showing its component parts and adsorption onto a host cell. **(b)** Penetration of a host cell by the bacteriophage.

nutrients are added as an **enrichment procedure** (Exercise 12). After incubation, the bacteriophage can be isolated by centrifugation of the enrichment media and membrane filtration. **Filtration** has been used for removing microbes from liquids for purposes of sterilization since 1884, when (Pasteur's associate) Charles Chamberland designed the first filter candle. Today most filtration of viruses is done using membrane filters with pore sizes (usually 0.45 μm) that physically exclude bacteria from the filtrate (Appendix F).

You will measure the viral activity in your sample by performing sequential dilutions of the viral preparation and assaying for the presence of viruses. In the

broth-clearing assay, the **endpoint** is the highest dilution (smallest amount of virus) producing lysis of bacteria and clearing of the broth. The **titer,** or concentration, that results in a recognizable effect is the reciprocal of the endpoint. In the plaque-forming method, the titer is determined by counting plaques. Each plaque theoretically corresponds to a single infective virus in the initial suspension. Some plaques may arise from more than one virus particle, and some virus particles may not be infectious. Therefore, the titer is determined by counting the number of **plaque-forming units (p.f.u.).** The titer, plaque-forming units per milliliter, is determined by counting the number of plaques and multiplying by the reciprocal of the dilution. For example, 32 plaques in a $1:10^4$ dilution is equal to 32×10^4 or 3.2×10^5 p.f.u.

In this exercise, we will determine the viral activity by a plaque-forming assay and a broth-clearing assay.

Materials

Isolation

Use flies or sewage

Flies as source:
Houseflies (fresh) (20 to 25); bring your own
(Figure 37.2)
Trypticase soy broth (20 ml)
Mortar and pestle

Sewage as source:
Raw sewage (45 ml)
$10 \times$ nutrient broth (5 ml)
50-ml graduated cylinder

Sterile 125-ml Erlenmeyer flask

Centrifuge tubes

Screw-capped tube

Sterile membrane filter (0.45 μm)

Sterile membrane filter apparatus

Titration

Tubes containing trypticase soy broth (6)

Petri plates containing nutrient agar (6)

Tubes containing 3 ml melted soft trypticase soy agar (0.7% agar) (6)

1-ml sterile pipettes (7)

Culture

Escherichia coli broth

Techniques Required

Exercises 10 and 11; Appendices A, B, and F.

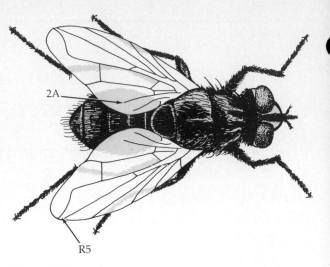

Figure 37.2 _____

Houseflies may harbor bacteriophages. To identify a housefly, observe that the 2A vein doesn't reach the wing margin, and the R5 cell narrows.

Procedure

> ☣ **Handle the flies and sewage carefully; potentially harmful microbes or chemicals may be present.**

Isolation

1. Follow enrichment procedure a or b.
   a. Add the flies to half of the trypticase soy broth in the mortar, and grind with the pestle to a fine pulp. Transfer this mixture to a sterile flask, and add the remainder of the broth and 2 ml *E. coli*. Wash the mortar and pestle in disinfectant and rinse with water.
   b. Add 45 ml sewage to 5 ml $10 \times$ nutrient broth in a sterile flask. Add 5 ml *E. coli* broth. Mix gently. Why is the broth ten times more concentrated than normal? _____
2. Incubate the enrichment for 24 hours at 35°C.
3. Decant 10 ml of the enrichment into a centrifuge tube. Place the tube in a centrifuge.

> ⚠ **Balance the centrifuge with a similar tube or a tube containing 10 ml water.**

Centrifuge at 2,500 rpm for 10 minutes to remove most bacteria and solid materials.
4. Filter the supernatant through a membrane filter. (Refer to Appendix F.) Decant the clear liquid into a screw-capped tube. How does filtration separate viruses from bacteria? _____

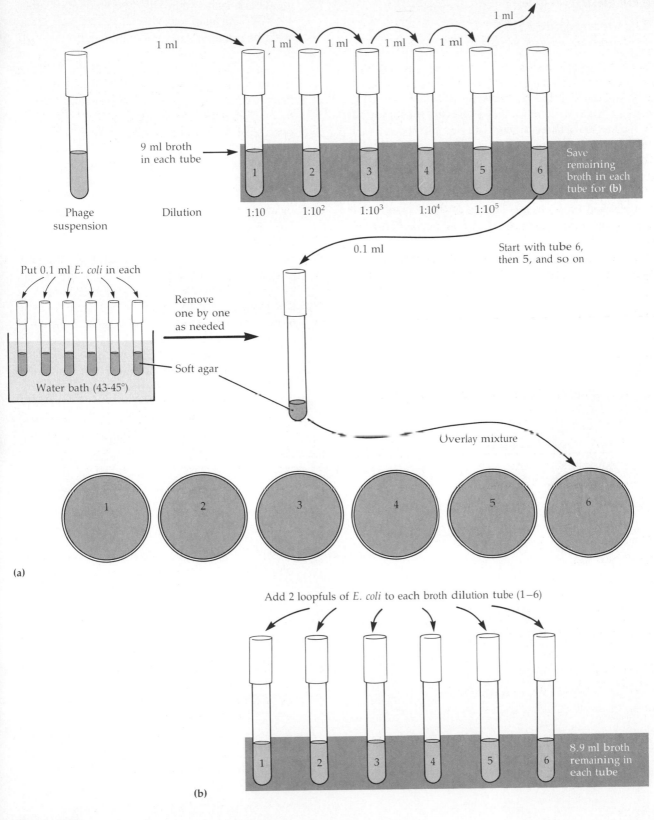

Figure 37.3 _____

Procedure for titration of bacteriophages. **(a)** Plaque-forming assay. **(b)** Broth-clearing assay.

Procedure

Titration (Figure 37.3)

Read carefully before proceeding.

1. Label the plates and the broth tubes "1" through "6."

2. Aseptically add 1 ml phage suspension (from step 4, above) to tube 1. Mix by carefully aspirating up and down three times with the pipette. Using a different pipette, transfer 1 ml to the second tube, mix well, and then put 1 ml into the third tube. Why is the pipette changed? _____
   Continue until the fifth tube. After mixing this tube, discard 1 ml into a container of disinfectant. What dilution exists in each tube? (See Figure 37.3.)

	Tube 1	Tube 2	Tube 3
Dilutions	_____	_____	_____
	Tube 4	Tube 5	Tube 6
	_____	_____	_____

   What is the purpose of tube 6? _____

3. Add 0.1 ml *E. coli* to the soft agar tubes and place them back in the waterbath. Keep the waterbath at 43° to 45°C at all times. Why? _____

4. With the remaining pipette, start with broth tube 6 and aseptically transfer 0.1 ml from tube 6 to the soft agar tube. Mix by swirling, and quickly pour the inoculated soft agar evenly over the surface of Petri plate 6. Then, using the same pipette, transfer 0.1 ml from tube 5 to a soft agar tube, mix, and pour over plate 5. Continue until you have completed tube 1.

5. After completing step 4, add 2 loopfuls of *E. coli* to each of the remaining broth tubes and mix. Incubate at 35°C. Observe in a few hours.

6. Incubate the plates at 35°C until plaques develop. The plaques may be visible within hours. (See color plate VII.1.)

7. Select a plate with between 25 and 250 plaques. Count the number of plaques and multiply by 10; then multiply by the reciprocal of the dilution of that plate to give the number of plaque-forming units (p.f.u.) per milliliter. Why is the number of plaques multiplied by 10? _____
   Record your results. In the broth tubes, record the highest dilution that was clear as the endpoint. The titer is the reciprocal of the endpoint.

EXERCISE 37

Isolation and Titration of Bacteriophages

Name _____

Date _____

Lab Section _____

Purpose _____

_____

_____

Data

Plaque-Forming Assay

Choose one plate with 25 to 250 plaques:

Draw what you observed.

    Number of plaques = _____

    Dilution used for that plate = _____

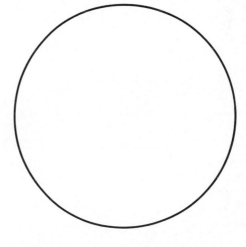

Broth-Clearing Assay

Indicate whether each tube was turbid or clear.

Incubated _____ hours.

Tube	Turbid or clear	Dilution
1		
2		
3		
4		
5		
6		

Interpretations

Plaque-Forming Assay

p.f.u./ml = _____
Show your calculations:

Broth-Clearing Assay

What was the endpoint? _____

What was the titer? _____

Questions

1. Are any contaminating bacteria present? How can you tell? _____

    _____

2. Why did you add *Escherichia coli* to sewage, which is full of bacteria? _____

    _____

    _____

3. Are all the plaques the same size? Briefly explain why or why not. _____

    _____

    _____

4. What does the endpoint represent in a broth-clearing assay? _____

    _____

    _____

    _____

5. Which of these assays is more accurate? Briefly explain. _____

    _____

    _____

6. How would you develop a pure culture of a phage? _____

    _____

    _____

7. If there were no plaques on your plates, offer an explanation. How would you explain turbidity in all of the tubes

    in a broth-clearing assay? _____

    _____

    _____

8. How would you isolate a bacteriophage for a species of *Bacillus*? _____

    _____

    _____

Plant Viruses

*Smoking was definitely dangerous to your
health in seventeenth-century Russia. Czar
Michael Federovitch executed anyone on
whom tobacco was found. But Czar Alexei
Mikhailovitch was easier on smokers; he
merely tortured them until they told
who their suppliers were.*

ANONYMOUS

Objectives

After completing this exercise you should be able to:

1. Isolate a plant virus.
2. Describe the cultivation of a plant virus.
3. Determine the host range of a plant virus.

Background

Plant viruses are very important economically in that they are the second leading cause of plant disease. (Fungi are the major cause of plant disease.) In order for most plant viruses to infect a plant, the virus must enter through an abrasion in the leaf or stem. Insects called **vectors** carry viruses from one plant to another. Most viruses cause either a **localized infection,** in which the leaf will have necrotic lesions (brown plaques), or a **systemic infection,** in which the infection will run throughout the entire plant. One of the most studied viruses is tobacco mosaic virus (TMV), which was the first virus to be purified and crystallized.

In this experiment, we will attempt to isolate TMV from various tobacco products and determine its host range.

Materials

Tobacco plant

Tomato plant

Bean plant (*Chenopodium*)

Tobacco products of various types (bring some, if possible)

Mortar and pestle

Fine sand or carborundum

Wash bottle of water

Paper labels or labeling tape

Cotton

Sterile water

Techniques Required

None.

Procedure

1. Select a plant and label the pot with your name and lab section.
2. Place labels around the petioles (see Figure 30.1) of several leaves, with names corresponding to the various tobacco products available. Label one petiole "C" for control.
3. *Wash your hands carefully before and after each inoculation.*
4. Spray the control leaf with a small amount of water and dust the leaf with carborundum or sand.

> ⚠ **Be careful not to inhale carborundum dust. It is a lung irritant.**

Gently rub the leaf with a cotton ball dampened in sterile water. Wash the leaf with a wash bottle of water. What is the purpose of this leaf? _____

5. Select a tobacco product, place a "pinch" of it into a mortar, and add sterile water. Grind into a slurry with a pestle. Spray one leaf with water, dust with carborundum or sand, and gently rub with a cotton ball soaked with the tobacco slurry. Wash the leaf

with water. Record the name of the tobacco source.

6. Using the same tobacco product, repeat steps 1 through 5 on a different species of plant, cleaning the mortar and pestle with soap and water between each product.

7. Replace the plant in the rack.

8. Check the plant at each laboratory period for up to 3 to 4 weeks for brown plaques (localized infection) or wilting (systemic infection). Discard plants in the To Be Autoclaved area. (See color plate VII.2.)

EXERCISE 38

Plant Viruses

Name _____

Date _____

Lab Section _____

Purpose _____

_____

_____

Data

Obtain data from your classmates for the other plant species and tobacco products.

Identify Plant Species and Tobacco Product Used		Appearance of Leaves							
		Lab Period							
		1	2	3	4	5	6	7	8
1.	Control								
	Test								
2.	Control								
	Test								
3.	Control								
	Test								
4.	Control								
	Test								
5.	Control								
	Test								
6.	Control								
	Test								

Draw a plant showing locations of the labeled leaves and any signs of infection.

Conclusions

What can you conclude about the host range of TMV? _____

_____

_____

_____

Questions

1. Did systemic disease occur? How can you tell? _____

_____

2. How do insects act as vectors of plant viruses? _____

_____

_____

3. If the control leaf was damaged, what happened? _____

_____

4. Design an experiment to prove that damage to the test leaves was due to a viral infection. _____

_____

_____

_____

_____

5. A reservoir of infection is a continual source of that infection. What is a likely reservoir for TMV? _____

_____

PART 10

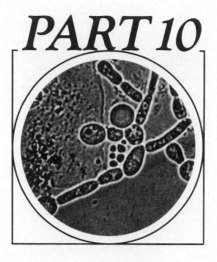

Interaction of Microbe and Host

_____ EXERCISES _____

39   Epidemiology   *40   Koch's Postulates*

In 1883, after having determined the number of microorganisms in the environment, Robert Koch asked:

> *What significance do these findings have? Can it be stated that air, water, and soil contain a certain number of microorganisms and yet are without significance to health?**

Koch felt that the causes, or **etiologic agents,** of infectious diseases could be found in the study of microorganisms. He proved that anthrax and tuberculosis were caused by specific bacteria, and his work provided the framework for the study of the etiology of any infectious disease.

Today we refer to Koch's protocols for identifying etiologic agents as **Koch's postulates** (see the figure and Exercise 40). The study of how and when diseases occur is called **epidemiology** (Exercise 39).

*Quoted in R. N. Coetsch, ed. *Microbiology: Historical Contributions from 1776 to 1908.* New Brunswick, NJ: Rutgers University Press, 1960, p. 130.

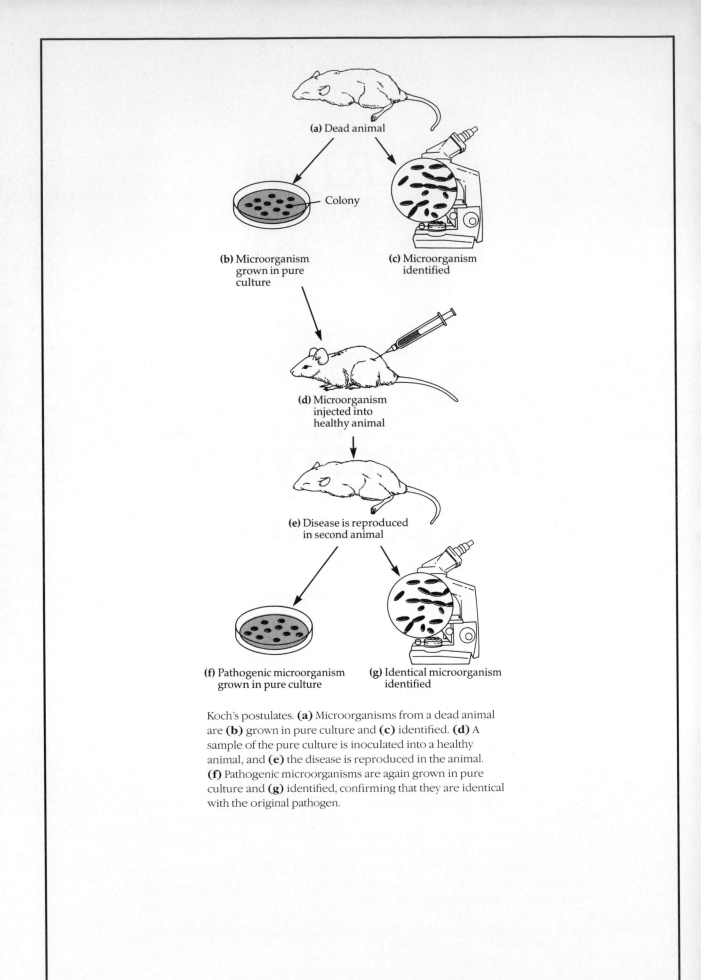

(a) Dead animal

Colony

(b) Microorganism grown in pure culture

(c) Microorganism identified

(d) Microorganism injected into healthy animal

(e) Disease is reproduced in second animal

(f) Pathogenic microorganism grown in pure culture

(g) Identical microorganism identified

Koch's postulates. **(a)** Microorganisms from a dead animal are **(b)** grown in pure culture and **(c)** identified. **(d)** A sample of the pure culture is inoculated into a healthy animal, and **(e)** the disease is reproduced in the animal. **(f)** Pathogenic microorganisms are again grown in pure culture and **(g)** identified, confirming that they are identical with the original pathogen.

EXERCISE 39

Epidemiology

Objectives

After completing this exercise, you should be able to:

1. Define the following terms: epidemiology, epidemic, reservoir, and carrier.
2. Describe three methods of transmission.
3. Determine the source of a simulated epidemic.

Background

In every infectious disease, the disease-producing microorganism, the **pathogen,** must come in contact with the **host,** the organism that harbors the pathogen. **Communicable diseases** can be spread either directly or indirectly from one host to another. Some microorganisms cause disease only if the body is weakened or if a predisposing event such as a wound allows them to enter the body. Such diseases are called **noncommunicable diseases;** that is, they cannot be transmitted from one host to another. The science that deals with when and where diseases occur and how they are transmitted in the human population is called **epidemiology. Sporadic diseases** are those that occur occasionally in a population; an example is polio. **Endemic diseases** such as pneumonia are constantly present in the population. When many people in a given area acquire the disease in a relatively short period of time, it is referred to as an **epidemic disease.** Influenza often achieves epidemic status.

Diseases can be transmitted by **direct contact** between hosts. **Droplet infection,** when microorganisms are carried on liquid drops from a cough or sneeze, is a method of direct contact. Diseases can also be transmitted by contact with contaminated inanimate objects, or **fomites.** Drinking glasses, bedding, and towels are examples of fomites that can be contaminated with pathogens from feces, sputum, or pus.

Some diseases are transmitted from one host to another by vectors. **Vectors** are insects and other arthropods that carry pathogens. In **mechanical transmission,** insects carry a pathogen on their feet and may transfer the pathogen to a person's food. For example, houseflies may transmit typhoid fever from the feces of an infected person to food. Transmission of a disease by an arthropod's bite is called **biological transmission.** An arthropod ingests a pathogen while biting an infected host and then transfers the pathogen to a healthy person in its feces or saliva.

The continual source of an infection is called the **reservoir.** Humans who harbor pathogens but who do not exhibit any signs of disease are called **carriers.**

An **epidemiologist** compiles data on the incidence of a disease and its method of transmission and tries to locate the source of infection in order to decrease the incidence.

In this exercise, an epidemic will be simulated. You will be the epidemiologist, who, by deductive reasoning and with luck, determines the source of the epidemic.

Materials

Petri plate containing nutrient agar

Small plastic sandwich bag

1 unknown swab per student (one swab has *Serratia marcescens* on it)

Techniques Required

Exercise 9.

Procedure

1. Divide the Petri plate into five sectors labeled "1" to "5."

Carefully read steps 2 through 4 before proceeding.

2. Put the plastic bag on your left hand. Carefully unwrap your swab without touching the cotton. Holding the swab with your right hand, rub it on the palm of your left hand (i.e., on the plastic bag). Discard the swab in a container of disinfectant.
3. Shake hands (using your left hand) with a classmate when the instructor gives the signal. Shake hands so your fingers touch the other's palm and vice versa. After shaking hands, touch your fingers to the first sector of the nutrient agar. Record the person's name and swab number.
4. Repeat step 3, shaking hands with four other classmates. Remember to touch your fingers to the corresponding sector of the nutrient agar after each handshake. Keep good records.
5. Discard the plastic bag in the To Be Autoclaved basket. Incubate the plate inverted at room temperature.
6. At the next lab period, deductively try to determine who had the contaminated fomite. (See color plate V.1.)

EXERCISE 39

Epidemiology

Name _____

Date _____

Lab Section _____

Purpose _____

_____

_____

Data

Results

Sector	Student's Name	Swab Number	Appearance of Colonies on Nutrient Agar
1.			
2.			
3.			
4.			
5.			

Conclusions

1. Who had the contaminated swab? _____

2. What number was the contaminated swab? _____ Explain how you arrived at your conclusion. _____

_____

3. Diagram the path of the epidemic.

Questions

1. Could you be the "infected" individual and not have growth on your plate? Explain. _____

   _____

   _____

   _____

2. Do all people who contact an infected individual acquire the disease? _____

   _____

   _____

   _____

3. When does an epidemic stop? _____

   _____

   _____

   _____

4. Are any bacteria other than *Serratia* growing on the plates? How can you tell? _____

   _____

   _____

5. What was the method of transmission of the "disease" in this experiment? _____

   _____

   _____

Koch's Postulates

Objectives

After completing this exercise you should be able to:

1. Define the following terms: etiologic agent, pathogenicity, and virulence.
2. List and explain Koch's postulates.

Background

The **etiologic agent** is the cause of a disease. Microorganisms are the etiologic agents of a wide variety of infectious diseases in all forms of life. Microbes that cause diseases are called pathogens, and the process of disease initiation and progress is called **pathogenesis.** The interaction between the microbe and host is complex. Whether or not a disease occurs depends on our vulnerability or **susceptibility** to the pathogen and on the virulence of the pathogen. **Virulence** is the degree of pathogenicity. Factors influencing virulence include the number of pathogens, the portal of entry into the host, and toxin production.

The actual cause of many diseases is hard to determine. Although many organisms can be isolated from a diseased tissue, their presence does not prove that any or all of them caused the disease. A microbe may be a secondary invader or part of the normal flora or transient flora of that area. While working with anthrax and tuberculosis, Robert Koch established four criteria, now called **Koch's postulates,** to help identify a particular organism as the causative agent for a particular disease (see the figure on p. 236). Koch's postulates are the following:

1. The same organism must be present in every case of the disease.
2. The organism must be isolated from the diseased tissue and grown in pure culture in the laboratory.
3. The organism from the pure culture must cause the disease when inoculated into healthy, susceptible laboratory animals.
4. The organism must again be isolated—this time from the inoculated animals—and must be shown to be the same pathogen as the original organism.

These criteria are used by most investigators, but they cannot be applied to all infectious diseases. For example, viruses cannot be cultured on artificial media, and they are not readily observable in a host. Moreover, many viruses cause diseases only in humans and not in laboratory animals.

In this exercise, we will demonstrate Koch's postulates with one of two different bacteria:

Vibrio anguillarium isolated from Chinook salmon

Erwinia carotovora isolated from soft rot of carrots

Materials
Vibrio: Goldfish
Petri plate containing nutrient agar

Petri plate containing salt–starch agar

Scalpel and scissors

Syringe containing broth

Goldfish (2)

Extra fish tank (for inoculated fish)

Demonstration
Dissected control fish

Erwinia: Carrot Soft Rot
Petri plate containing nutrient agar

Carrot

Scalpel or razor blade, potato peeler

Forceps

Alcohol

Disinfectant (bleach)

Sterile water

Sterile Petri plate with filter paper

Cultures
Vibrio: Goldfish
Syringe containing an 18-hour culture of *Vibrio anguillarium*

Erwinia: Carrot Soft Rot
Erwinia carotovora

Techniques Required

Exercises 5, 11, and 13.

Procedure

Vibrio: Goldfish*

1. Divide each Petri plate into three sectors; label them A, B, and C. Streak a drop of broth from the syringe onto sector A and a drop of *Vibrio* from the other syringe onto sector B of each plate. Incubate the plates inverted at room temperature until growth occurs, then store in the refrigerator.

2. Make a smear of *Vibrio* and heat fix; store in your drawer.

3. Remove one fish with a wet paper towel. Do not touch the fish directly. Why? _____

_____

Holding the fish on its back, inject it with *Vibrio* in front or behind the anal fin (Figure 40.1). Hold the syringe at about a 45° angle and inject about 0.05 ml. Put the fish into the tank for inoculated fish.

> ☣ **Do not recap the syringe after use. Place the syringe and cap in the appropriate container.**

4. Inject the remaining fish in the same manner with 0.05 ml broth. What is the purpose of this step?

_____

_____

Treat the fish with care. Return it to the tank for inoculated fish.

> ☣ **Discard the syringe without recapping.**

5. The inoculated fish should die within 24 to 48 hours. Once dead, place in a wet towel and keep in the refrigerator until you have time to necropsy the fish. *If your fish does not die, you have a new pet.* Provide some reasons for its survival.

6. Observe the viscera of the dissected control fish.

7. Work with the fish on a disinfectant-saturated towel. Disinfect your instruments by putting them in alcohol and burning off the alcohol immediately before and after use. Slit the belly open from anus to gills, then make horizontal cuts to expose the viscera (Figure 40.2). Examine and record your findings. Transfer a loopful of the peritoneum exudate (fluid or pus) to sector C on your plates stored previously in the refrigerator (step 1) and incubate at room temperature for 48 hours. Make a smear of exudate on the same slide you used in step 2. Discard the fish as instructed.

8. Gram stain the smears and examine the growth on the plates. If growth occurred on nutrient agar, a contaminant was present. Growth on the plates can also be Gram stained. Flood the salt–starch plate with Gram's iodine. Record your results.

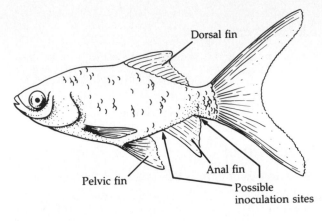

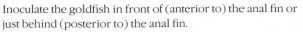

Figure 40.1 _____
Inoculate the goldfish in front of (anterior to) the anal fin or just behind (posterior to) the anal fin.

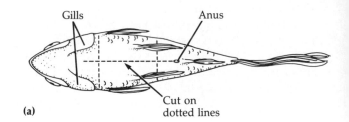

(a)

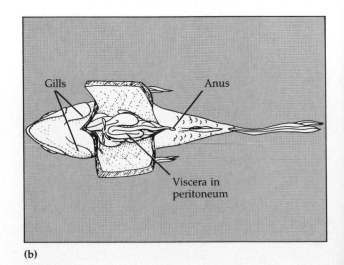

(b)

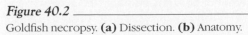

Figure 40.2 _____
Goldfish necropsy. **(a)** Dissection. **(b)** Anatomy.

---

*Adapted from W. Umbreit and E. Ordal. "Infection of Goldfish with *Vibrio anguillarium*." *ASM News* 38:93–98, 1972. With the permission of the American Society for Microbiology.

Erwinia: Carrot Soft Rot*

1. Wash the carrot well, and peel it to eliminate the outer surface; then dry, wash with disinfectant, and rinse with sterile water. Dip your scalpel in alcohol, burn off the alcohol, and cut the carrot into four cross-sectional slices 5 to 8 mm thick.
2. Put the four slices on filter paper in the bottom of a Petri plate.
3. Inoculate the center of three slices with a loopful of the *Erwinia* culture. Why not all four? _____
    _____

    Saturate the filter paper with sterile water. Incubate the plate right-side up at room temperature until

*Adapted from R. S. Hogue. "Demonstration of Koch's Postulates." *American Biology Teacher* 33:174–175, 1971.

soft rot appears. (See color plate VII.3.) More sterile water may be added if the disease process is slow.

4. Divide the nutrient agar plate in half. Inoculate one-half of the nutrient agar with the *Erwinia* broth, incubate inverted for 48 hours at room temperature, and then refrigerate. Make a smear of the *Erwinia*. Heat fix the smear and store in your drawer.
5. Streak an inoculum from the diseased carrot on the remaining half of the Petri plate. Incubate inverted at room temperature for 48 hours.
6. Make a smear from the diseased carrot. Gram stain both smears and record your observations.
7. Observe the nutrient agar plate and record your results. Prepare a Gram stain from the nutrient agar cultures if time permits.

EXERCISE 40

Koch's Postulates

Name _____

Date _____

Lab Section _____

Purpose _____

_____

_____

Vibrio: Goldfish

Gram Stains

Vibrio anguillarium
(before inoculation)
Morphology: _____

Gram reaction: _____

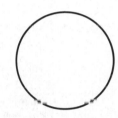

Peritoneum fluid from dead fish
Morphology: _____

Gram reaction: _____

Cultures

Describe the growth seen.

Source	Nutrient Agar	Gram Stain	Salt-Starch Agar	Starch Hydrolysis?
Saline				
Vibrio preinoculation				
Peritoneum fluid				

Necropsy observations:
Describe your findings. (Compare your fish with the dissected demonstration.) _____

_____

_____

Pus? _____ Hemorrhage? _____

Conclusions _____

_____

_____

_____

_____

Questions

1. Do you think this bacterium produces a toxin? _____

    _____

    _____

2. Why was salt–starch agar used to reisolate the *Vibrio*? _____

    _____

    _____

3. If your broth-inoculated fish died, what would this suggest about the method of disease transmission? _____

    _____

    _____

    _____

4. What evidence would indicate that the microbe obtained from the dead fish was *Vibrio anguillarium?* _____

    _____

    _____

    _____

5. Were Koch's postulates fulfilled? _____ Explain. _____

    _____

    _____

Erwinia: Carrot Soft Rot

Gram Stains

Erwinia carotovora
Morphology: _____

Gram reaction: _____

Soft Rot
Morphology: _____

Gram reaction: _____

Cultures

Inoculum	Colony Description	Gram Stain
Erwinia		
Soft rot		

Describe the appearance of the carrots. _____

_____

_____

_____

Conclusions _____

_____

_____

_____

Questions

1. Why is the disease called soft rot? _____

_____

2. Why will the addition of water speed the disease process? _____

_____

_____

3. Were Koch's postulates fulfilled? _____

_____

_____

Summary Questions

1. Give two examples (other than those used in this exercise) of instances where Koch's postulates have not been

fulfilled. _____

_____

_____

2. What parts of Koch's postulates did these experiments fulfill? _____

_____

_____

_____

3. What is the normal method of transmission of the following:

    a. *Vibrio anguillarium?* _____

    b. *Erwinia carotovora?* _____

PART 11

Immunology

EXERCISES

In this part, we will examine our defenses against invasion by microorganisms. In 1883, Elie Metchnikoff began his investigations into the body's defenses with this entry into his diary:

> *These wandering cells in the body of the larva of a starfish, these cells eat food, they gobble up carmine granules — but they must eat up microbes too! Of course — the wandering cells are what protect the starfish from microbes! Our wandering cells, the white cells of blood — they must be what protects us from invading germs.** *

These white blood cells (see the figure) and certain chemicals are considered agents of nonspecific immunity because they will combat any microorganism that invades the body. Factors involved in nonspecific defenses is the topic of Exercise 41.

The role of specific defenses — or **immunity** — was demonstrated when Emil von Behring developed what he called diphtheria "antitoxin" and, in 1894, successfully treated humans with it. Immunity involves the production of specific

---

*Quoted in P. DeKruif. *Microbe Hunters*. New York: Harcourt, Brace, & World, 1953, p. 195.

proteins called **antibodies,** which are directed against distinct microorganisms. If microorganism A invades the body, antibodies are produced against A; if microorganism B invades the body, antibodies are produced against B. The antibodies produced against microorganism B will not react with microorganism A. The chemical substance that induces antibody production is called an **antigen.** In the examples just mentioned, microorganisms A and B possess antigens on their surfaces.

Specific and nonspecific immunity involves blood. **Blood** consists of a fluid called **plasma** and formed elements: cells and cell fragments. The cells that are of immunologic importance are the white blood cells. **Serum** is the fluid portion that remains after blood has clotted. We will study the techniques and mechanisms of **serology,** or antigen–antibody reactions in vitro, in Exercises 42 through 44.

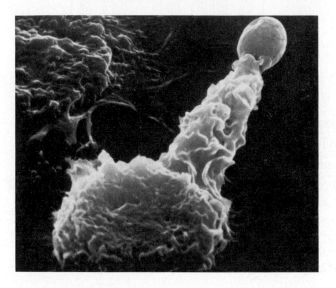

Early stage of phagocytosis by an alveolar macrophage (a type of macrophage found in the alveoli of the lungs). The "wrinkled" macrophage has contacted and adhered to a smooth, roughly spherical yeast cell by means of a pseudopod (3100 ×).

Nonspecific Resistance

Objectives

After completing this exercise you should be able to:

1. Differentiate between specific and nonspecific immunity.
2. List and discuss the functions of at least three chemicals involved in nonspecific resistance.
3. Use a spectrophotometer.
4. Determine the effects of normal serum bactericidins.

Background

Our ability to ward off diseases through our body's defenses is called **resistance.** Resistance can be divided into two kinds: specific and nonspecific. Specific resistance, or **immunity,** is the defense against a specific microorganism (Exercise 43). **Nonspecific resistance** refers to all our defenses that protect us from invasion by any microorganism. Physical barriers, such as the skin and mucous membranes, are part of nonspecific resistance. Inflammation and phagocytosis play major roles in nonspecific resistance, as do chemicals produced by the body. As early as 1888, Metchnikoff noted that bacteria "show degenerative changes" when inoculated into samples of mammalian blood.

Lysozyme and complement are examples of chemicals involved in nonspecific resistance. These proteins are collectively called **bactericidins. Lysozyme** is an enzyme found in body fluids that is capable of breaking down the cell walls of gram-positive bacteria and a few gram-negative bacteria. **Complement** is a group of proteins found in normal serum that are involved in phagocytosis and lysis of bacteria. Complement can be activated by bacterial cell wall polysaccharides and antigen–antibody reactions (Exercise 43).

In this exercise, we will examine chemicals involved in nonspecific resistance.

Materials

Lysozyme Activity

Lysozyme buffer, 4.5 ml

Egg-white lysozyme, $1:10^5$ dilution, 5 ml

Spectrophotometer tubes (2)

Pipettes, 1.0 ml and 5.0 ml (one of each)

Spectrophotometer

Normal Serum Bactericidins

Nutrient agar, melted and cooled to 45°C

Sterile Petri plates (9)

Sterile pipettes, 1.0 ml (10)

Sterile 99-ml dilution blanks (3)

Sterile serological tubes (9)

Serological test-tube rack

0.85% saline solution

Normal animal serum

Animal serum heated to 56°C for 30 minutes

Cultures (as assigned)

Micrococcus lysodeikticus suspension, 5 ml

Escherichia coli

Staphylococcus aureus

Techniques Required

Exercises 10 and 11; Appendices A, B, C, and D.

Procedure

Work only with your own tears or saliva!

Lysozyme Activity

1. Collect tears or saliva in a Petri plate, as explained by your instructor.
2. Prepare a 1:10 dilution of tears or saliva by adding 4.5 ml lysozyme buffer to 0.5 ml tears or saliva. What is the purpose of the buffer? _____

   _____

3. Mix 5 ml of the tear or saliva preparation with a 5-ml *Micrococcus lysodeikticus* suspension in a spectrophotometer tube. Carefully pipette the contents of the tube up and down three times to mix.
4. Record the absorbance at 540 nm at 30-second intervals for 5 minutes, and at 5-minute intervals for the next 15 minutes.
5. Repeat steps 2 and 3 using egg-white lysozyme.
6. Plot your data in the Laboratory Report.

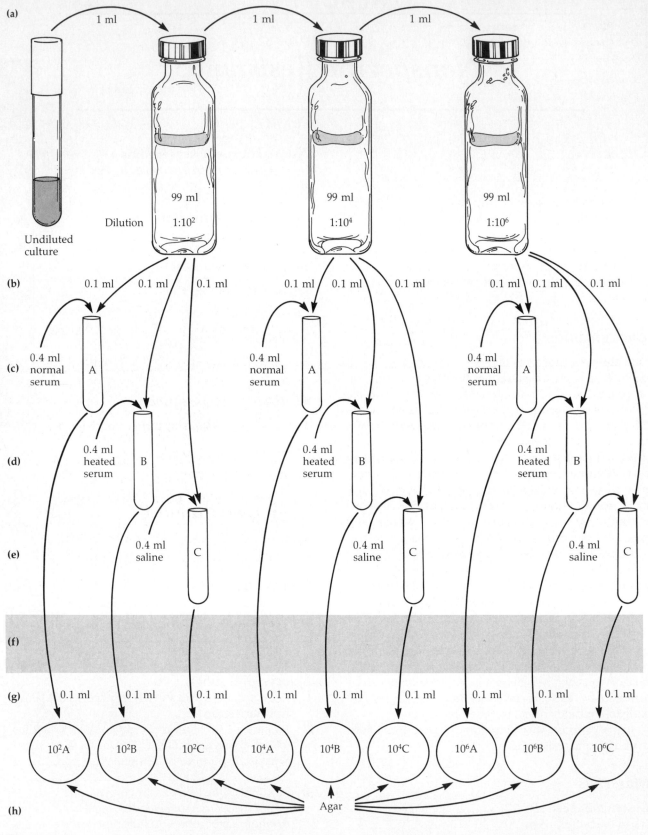

(a)

1 ml   1 ml   1 ml

Undiluted
culture

Dilution   99 ml   99 ml   99 ml
1:10²   1:10⁴   1:10⁶

(b)

0.1 ml   0.1 ml   0.1 ml   0.1 ml   0.1 ml   0.1 ml   0.1 ml   0.1 ml   0.1 ml

(c) 0.4 ml   0.4 ml   0.4 ml
normal   A   normal   A   normal   A
serum   serum   serum

(d)   0.4 ml   0.4 ml   0.4 ml
heated   B   heated   B   heated   B
serum   serum   serum

(e)   0.4 ml   0.4 ml   0.4 ml
saline   C   saline   C   saline   C

(f)

(g) 0.1 ml   0.1 ml   0.1 ml   0.1 ml   0.1 ml   0.1 ml   0.1 ml   0.1 ml   0.1 ml

10²A   10²B   10²C   10⁴A   10⁴B   10⁴C   10⁶A   10⁶B   10⁶C

(h)   Agar

Figure 41.1 _____

Determining the bactericidal activity of serum.

> **Place the Petri plate, dilution tube, and spectrophotometer tube in disinfectant.**

Normal Serum Bactericidins
(Figure 41.1)
Work in groups as assigned.

1. Aseptically prepare $1:10^2$, $1:10^4$, and $1:10^6$ dilutions of the culture *(S. aureus* or *E. coli)* assigned to you (Figure 41.1*a*).
2. Label three sterile serological tubes "6A," "6B," and "6C." Aseptically transfer 0.1 ml of culture from the highest dilution (1: _____) to each tube (Figure 41.1*b*).
3. Repeat step 2 with the other dilutions using the same pipette. Label each tube appropriately. How many tubes do you have? _____
4. To each A tube, add 0.4 ml of unheated normal serum; to each B tube, add 0.4 ml of serum heated to 56°C for 30 minutes; to each C tube, add 0.4 ml of saline (Figure 41.1*c, d,* and *e*). What is the purpose of the C tubes? _____
5. Mix all tubes and incubate at 35°C for 1 hour (Figure 41.1*f*).
6. Label nine sterile Petri plates in the same manner as the tubes: "$1:10^2$ A," "$1:10^2$ B," "$1:10^2$ C," "$1:10^4$ A," "$1:10^4$ B," and so on.
7. Remove 0.1 ml from the A tube of the highest dilution, and place it in the appropriate Petri plate. Using the same pipette, transfer 0.1 ml from the $1:10^4$ A tube to the corresponding Petri plate. Repeat this procedure with the $1:10^2$ A tube (Fig-

ure 41.1*g*). Why are the samples transferred from the highest dilution first? _____

8. Using another pipette, repeat step 6 with the B tubes. With your remaining pipette, repeat step 6 with the C tubes.
9. Pour melted, cooled nutrient agar into each plate to a depth of approximately 5 mm. Gently swirl to mix the contents and allow the agar to solidify.
10. Incubate the plates for 24 to 48 hours at 35°C. Record the number of colonies on each plate. Plates with fewer than 25 colonies should be reported as TFTC (too few to count) and those with more than 250 colonies as TNTC (too numerous to count).
11. Calculate the number of bacteria per milliliter for the control. Choose one plate with between 25 and 250 colonies:

$$\frac{\begin{matrix} \text{Number} \\ \text{of} \\ \text{colonies} \end{matrix} \times \begin{matrix} \text{Reciprocal} \\ \text{of} \\ \text{dilution} \end{matrix}}{0.1 \text{ ml}} = \text{Bacteria/ml}$$

Calculate the number of bacteria per milliliter for the unheated and heated sera.
12. Calculate the percent of increase or decrease in bacterial numbers due to exposure to heated and unheated serum as compared to the control.

$$\frac{\text{Control} - \text{Experiment}}{\text{Control}} \times 100 = \%$$

13. Compare your data with those of a group using the other bacterial species.

EXERCISE 41

Nonspecific Resistance

Name _____

Date _____

Lab Section _____

Purpose _____

_____

_____

Data

Lysozyme Activity

Time	Absorbance		
	Saliva	Tears	Egg-white Lysozyme
30 sec			
60 sec			
120 sec			
180 sec			
240 sec			
5 min			
10 min			
15 min			
20 min			

Plot your data for lysozyme activity on the graph paper on the next page. Absorbance is marked on the Y-axis and time on the X-axis. Make one line for tears or saliva and another line for egg-white lysozyme.

	Number of colonies			Bacteria/ml
	$1:10^2$	$1:10^4$	$1:10^6$	
Unheated serum				
Heated serum				
Control				

Your calculations:

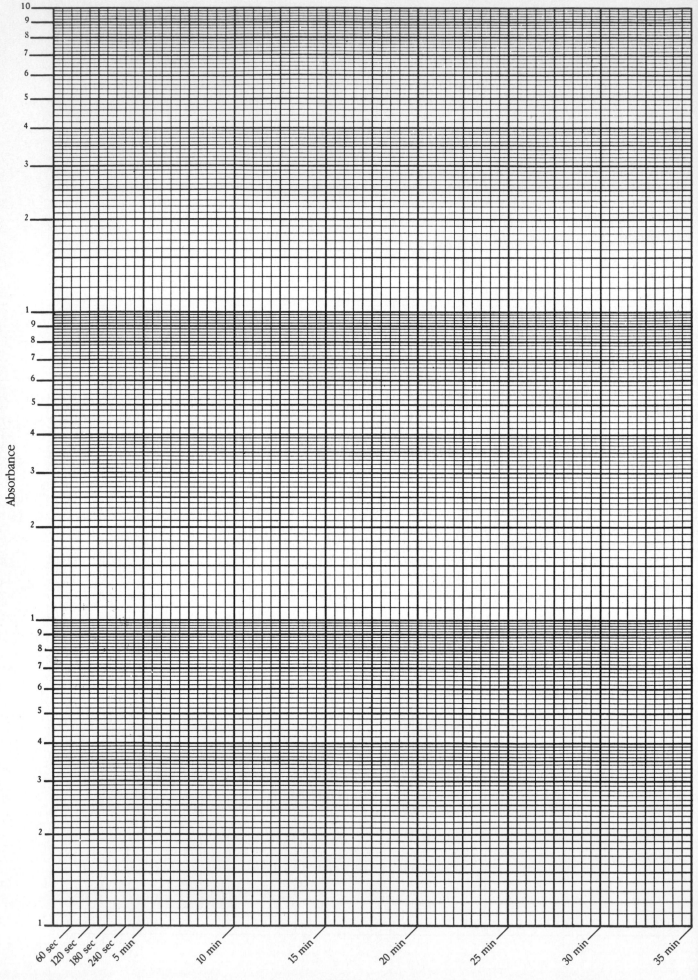

Conclusions

What effect did lysozyme have on the bacterial suspension? _____

_____

	% increase or decrease	
	S. aureus	*E. coli*
Unheated serum		
Heated serum		

Your calculations:

Questions

1. Why are tears and saliva biohazards? _____

_____

2. Was lysozyme present in your tears or saliva? _____

How can you tell? _____

3. Did you test tears or saliva? _____ Compare your data with someone who tested

the other. Which secretion has more lysozyme activity? _____ How can you tell?

_____

4. Does normal serum have bactericidal properties? _____

5. What can you conclude regarding the effect of heat on the bactericidal properties of normal serum? _____

_____

_____

6. Design an experiment to determine whether plant fluids have bactericidal properties. _____

_____

_____

_____

Blood Group Determination: Slide Agglutination

Objectives

After completing this exercise you should be able to:

1. Compare and contrast the terms agglutination and hemagglutination.
2. Determine ABO and Rh blood types.
3. Determine compatible transfusions.

Background

Agglutination reactions occur between high-molecular-weight, particulate antigens and antibodies. Since many antigens are on cells, agglutination reactions lead to the clumping, or **agglutination,** of cells. When the cells involved are red blood cells, the reaction is called **hemagglutination.**

Hemagglutination reactions are used in the typing of blood. The presence or absence of two very similar carbohydrate antigens (designated A and B) located on the surface of red blood cells is determined using specific antisera (Figure 42.1). Hemagglutination occurs when anti-A antiserum is mixed with type A red blood cells. When anti-A antiserum is mixed with type B red blood cells, no hemagglutination occurs. People with type AB blood possess both A and B antigens on their red blood cells, and those with type O blood lack A and B antigens (see Figure 42.1).

Many other blood antigen series exist on human red blood cells. Another surface antigen on red blood cells is designated the Rh factor. The **Rh factor** is a complex of many antigens. The Rh factor that is used routinely in blood typing is the **Rh_0 antigen,** or **D antigen.** Individuals are Rh-positive when D antigen is present. The presence of the Rh factor is determined by a hemagglutination reaction between anti-D antiserum and red blood cells.

A person possesses antibodies to the alternate antigen. Thus, people of blood type A will have antibodies to the B antigen in their sera (see Figure 42.1). Rh-negative individuals do not naturally have anti-D antibodies in their sera. Anti-D antibodies are produced when red blood cells with D antigen are introduced into Rh-negative individuals.

The ABO and Rh systems place restrictions on how blood may be transfused from one person to another. An incompatible transfusion results when the antigens of the donor red blood cells react with the antibodies in the recipient's serum or induce the formation of antibodies.

A summary of the major characteristics of the ABO and Rh blood groups is presented in Table 42.1.

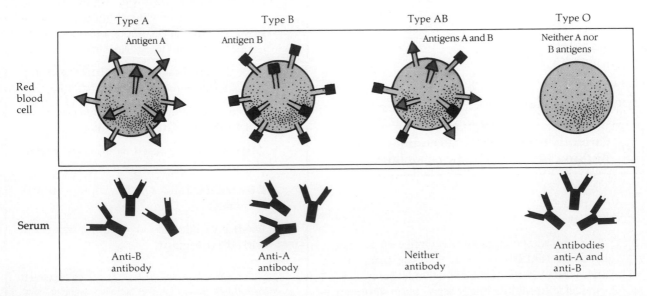

Figure 42.1

Relationship of antigens and antibodies involved in the ABO blood group system.

Table 42.1

The ABO and Rh Blood Group Systems

Characteristic	Blood Type					
	A	B	AB	O	Rh+	Rh−
Antigen present on the red blood cells	A	B	Both A and B	Neither A nor B	D	No D
Antibody normally present in the serum	anti-B	anti-A	Neither anti-A nor anti-B	Both anti-A and anti-B	No anti-D	No anti-D*
Serum causes agglutination of red blood cells of these types	B, AB	A, AB	None	A, B, AB	Neither Rh+ nor Rh−	Neither Rh+ nor Rh−*
Cells agglutinated by serum of these types	B, O	A, O	A, B, O	None	Neither Rh+ nor Rh−	Neither Rh+ nor Rh−
Percent occurrence in a mixed Caucasian population	41	10	4	45	85	15
Percent occurrence in a mixed Black population	27	20	7	46	85	15

*Anti-D antibodies are not naturally present in the serum of Rh− people. Anti-D antibodies can be produced upon exposure to the D antigen through blood transfusions or pregnancy.

Source: Adapted from G. J. Tortora, B. R. Funke, and C. L. Case. *Microbiology: An Introduction*, 4th ed. Redwood City, CA: Benjamin/Cummings, 1992.

Materials

Grease pencil

Cotton moistened with 70% ethyl alcohol

Sterile cotton ball and bandage

Sterile lancet

Anti-A, anti-B, and anti-D antisera

Glass slide (2)

Toothpicks

Techniques Required

> Carefully read the Safety Precautions for Working with Blood in the Laboratory.

Procedure

1. With a grease pencil, draw two circles on a clean glass slide, and label one "A" and the other "B." Draw a circle on the second slide and label it "D."
2. Disinfect your middle finger with cotton saturated in alcohol (Figure 42.2). Pierce the disinfected finger with a sterile lancet.

Safety Precautions for Working with Blood in the Laboratory

- Do not perform this lab experiment if you are sick.
- Work only with **your own** blood* or wear gloves.
- **Disinfect** your finger with 70% alcohol. Use any finger except your thumb. Unwrap a **new, sterile** lancet and pierce the disinfected finger with the sterile lancet.
- Place the used lancet **in disinfectant** in a container designated by the instructor.
- Stop the bleeding with a sterile cotton ball and **apply an adhesive bandage.**
- **Dispose** of the used cotton **in disinfectant.**
- **Discard** used slides and toothpicks **in disinfectant.**
- **Wash** any spilled blood from your work area **with disinfectant.**

*Blood obtained from blood banks is tested for hepatitis B virus and HIV. No test method can offer complete assurance that laboratory specimens do not contain these viruses.

3. Let a drop of blood fall into each circle. Stop the bleeding with a sterile cotton ball and apply an adhesive bandage.

4. To the A circle, add 1 drop of anti-A antiserum. Add 1 drop of anti-B antiserum to the B circle and 1 drop of anti-D antiserum to the D circle.

5. Mix each suspension with toothpicks. Use a different toothpick for each antiserum. Why? _____

_____

6. Observe for agglutination and determine your blood type. (See color plate IX.1.) A dissecting microscope may help you see hemagglutination (Appendix E).

> ☣ **Discard the slides, toothpicks, and lancet in the disinfectant.**

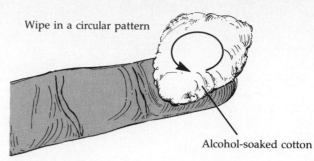

Wipe in a circular pattern

Alcohol-soaked cotton

Figure 42.2 _____

To disinfect skin, rub with 70% ethyl alcohol and wipe dry with a sterile cotton ball.

EXERCISE 42

Blood Group Determination: Slide Agglutination

Name _____

Date _____

Lab Section _____

Purpose _____

_____

_____

Data

Antiserum	Hemagglutination
Anti A	
Anti-B	
Anti-D	

Data Analysis

On the blackboard, tabulate the blood types for the class or laboratory section. Calculate the percentage of each blood type, and compare with the percent distribution given in Table 42.1.

Questions

1. What is your blood type? _____

2. What is the blood type of the person on your left? _____

    Is your blood compatible? _____ Explain briefly. _____

3. What is the blood type of the person on your right? _____

    Is your blood compatible? _____ Explain briefly. _____

4. What is hemolytic disease of the newborn? How does Rhogam® prevent it? _____

    _____

    _____

5. How can blood type O be considered the "universal donor"? _____

    _____

Agglutination Reactions: Microtiter Agglutination

Objectives

After completing this exercise you should be able to:

1. Define agglutination and titer.
2. Determine the titer of antibodies by the agglutination method.
3. Provide two applications for agglutination reactions.

Background

The outer surfaces of bacterial cells contain antigens that can be used in agglutination reactions. **Agglutination reactions** occur between *particulate* antigens — such as cell walls, flagella, or capsules *bound* to cells — and antibodies. **Agglutination,** or the clumping of bacteria by antibodies, is a useful laboratory diagnostic technique.

In a slide agglutination test (Exercise 42), an unknown bacterium is suspended in a saline solution on a slide and mixed with a drop of known antiserum. This test is done with different antisera on separate slides. The bacteria will agglutinate when mixed with antibodies produced against the same strain and species. A positive test can identify the bacterium.

An **agglutination titration** can estimate the concentration (*titer*) of antibody in serum. This test is used to determine whether a particular organism is causing the patient's symptoms. If an increase in titer is shown in successive daily tests, the patient probably has an infection caused by the organism used in the test.

In the titration, bacteria are mixed with dilutions of serum (e.g., 1:5, 1:10, 1:20, 1:40, and so on). The *endpoint* is the greatest dilution of serum showing an agglutination reaction. The *reciprocal* of this dilution is the titer. For example, in Figure 43.1, the titer is 40.

In this exercise, we will perform a test to diagnose typhoid fever, the **Widal test.**

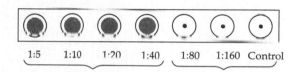

(a) Top view of wells

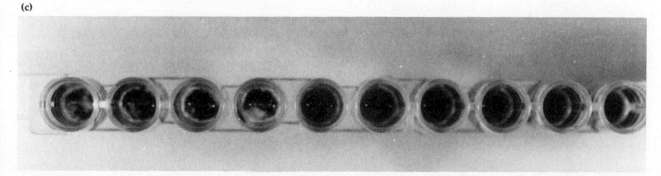

(b) Enlarged side view of wells

(c)

Figure 43.1 _____

(a) Each well in this microtiter plate contains, from left to right, only half the concentration of serum in the preceding well. Each well contains the same concentration of bacterial cells. **(b)** In a positive reaction, sufficient antibodies are present in the serum to link the antigens together, forming an antibody–antigen mat that sinks to the bottom of the well. In a negative reaction, insufficient antibodies are present to cause the linking of antigens. **(c)** Appearance of wells after refrigeration for 24 hours.

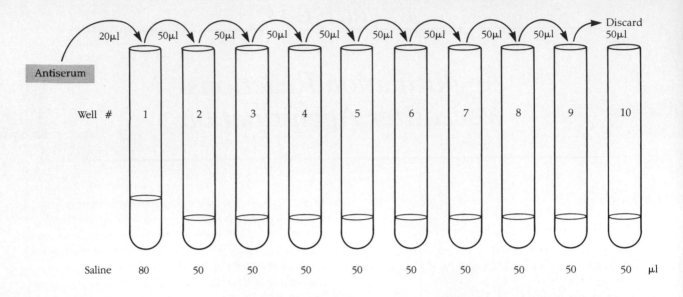

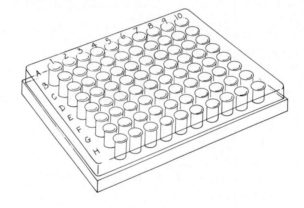

Figure 43.2 _____

Dilution of antiserum. Twenty µl of patient's serum is mixed with 80 µl of saline in well 1. Serial dilutions are made by transferring 50 µl from one well to the next, through well 9. Discard 50 µl from well 9.

> **☣ _Salmonella typhimurium_ is a causative agent of gastroenteritis. Be careful when handling this pathogenic bacterium.**

Materials

0.85% saline solution

Patient's serum

Microtitration plate

Micropipettes, 20 to 80 µl

Micropipette tips (13)

Culture

Salmonella typhimurium (heated for 30 minutes at 56°C in water bath)

Techniques Required

Appendices A and B.

Procedure

1. Label ten wells in the microtitration plate "1" through "10" (Figure 43.2).
2. Add 80 µl saline to the first well and 50 µl to each of the remaining wells.
3. Add 20 µl patient's serum to the first well. Mix up and down three times and, with a new pipette tip, transfer 50 µl to the second well. Change pipette tips, mix, and transfer 50 µl to the third well, and so on. Continue until you have reached the ninth well. Discard the 50 µl from that well, as shown in Figure 43.2. The patient's serum has now been diluted. The dilutions are as follows:

Well 1   1:5

Well 2   1:10

Well 3   1:20

Well 4   1: _____

Well 5   1: _____

Well 6   1: _____

Well 7   1: _____

Well 8   1: _____

Well 9   1: _____

Well 10   no serum

Complete the missing dilutions.

4. Carefully add 50 μl of the antigen to each well. What is the antigen? _____
   How does the addition of 50 μl of antigen affect the dilutions? _____

5. Shake the plate carefully in a horizontal direction to mix the contents in each well. Place the plate in a 35°C incubator for 60 minutes.

6. Refrigerate until the next laboratory period.

7. Observe the bottom of each well for agglutination. A dissecting microscope (Appendix E) may help you see agglutination. Which well serves as the control? _____
   _____

   What should occur in this well? _____
   _____

   Determine the endpoint and the titer.

EXERCISE 43

Agglutination Reactions: Microtiter Agglutination

Name _____

Date _____

Lab Section _____

Purpose _____

_____

_____

Data

Well #	Final Dilution	Agglutination
1		
2		
3		
4		
5		
6		
7		
8		
9		
10		

Diagram the appearance of the bottom of a positive and negative well:

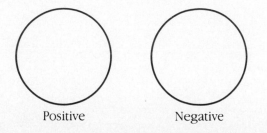

Positive   Negative

Questions

1. How did you determine the presence of agglutination? _____

_____

2. What was the endpoint? _____

3. What was the antibody titer? _____

4. What was the antiserum used in this experiment? _____

5. How would agglutination reactions help in locating the source of an epidemic? _____

_____

_____

6. The following antibody titers were obtained for three patients:

	Antibody Titer		
Patient	Day 1	Day 5	Day 12
A	1:128	1:128	1:128
B	1:128	1:256	1:512
C	0	0	0

What can you conclude about each of these patients? _____

_____

_____

Fluorescent-Antibody Technique

Objectives

After completing this exercise you should be able to:

1. Explain how fluorescent-antibody tests can be used to diagnose diseases.
2. Differentiate between direct and indirect fluorescent-antibody tests.
3. Compare and contrast direct fluorescent-antibody testing with simple staining and brightfield microscopy.

Background

Fluorescent-antibody (F.A.) techniques are used to identify microorganisms or to detect the presence of antibodies in a patient's serum. F.A. tests employ antibodies that have been combined **(conjugated)** with fluorescent dyes such as fluorescein isothiocyanate (FITC). **Fluorescent dyes** emit visible light when they absorb ultraviolet or near-ultraviolet wavelengths. A **fluorescent microscope** is a compound microscope with an ultraviolet or near-ultraviolet light source. In fluorescent microscopy, the object is seen as bright and luminescent against a dark background.

There are two types of fluorescent-antibody tests: direct and indirect. **Direct F.A. tests** are used to determine the identity of a microorganism (antigen). In this procedure, the specimen containing the antigen is fixed onto a slide. Fluorescent-labeled antibody is then added to the slide. After a brief incubation, the slide is washed and examined under a fluorescent microscope. The presence of fluorescing bacterial cells (Figure 44.1) is a positive test.

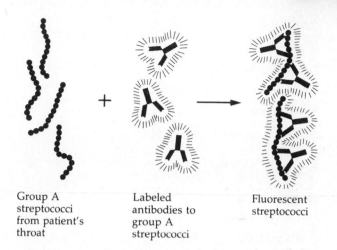

Group A
streptococci
from patient's
throat

Labeled
antibodies to
group A
streptococci

Fluorescent
streptococci

Figure 44.1 ⎯⎯⎯⎯⎯⎯⎯⎯⎯⎯⎯⎯⎯⎯⎯⎯⎯
Use of the direct fluorescent-antibody technique for the identification of group A streptococci.

Indirect F.A. tests are used to detect the presence of a specific antibody in a patient's serum. In this procedure, a known species (or serotype) of microorganism is fixed onto a slide. Serum is then added, and if antibody specific to the microorganism is present, it reacts with the antigens forming a bound complex. To observe this antigen–antibody complex, fluorescent-labeled antihuman gamma globulin *(antiantibody,* or *anti-HGG)* is added to the slide. After incubation and washing, the slide is observed for fluorescing antigen–antibody complexes under a fluorescent microscope (Figure 44.2).

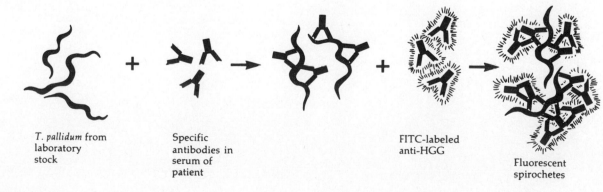

T. pallidum from
laboratory
stock

Specific
antibodies in
serum of
patient

FITC-labeled
anti-HGG

Fluorescent
spirochetes

Figure 44.2 ⎯⎯⎯
Use of the indirect fluorescent-antibody technique for the diagnosis of syphilis (fluorescent treponemal antibody test).

Materials

E. coli polyvalent fluorescent antibody

F.A. buffer

F.A. mounting medium

95% ethyl alcohol

Petri plate with filter paper liner (2)

Applicator sticks (2)

Gram staining reagents

Cover slips (2)

Cultures

Unknown bacterial cultures (2)

Techniques Required

Exercises 1 and 5.

Procedure

Gram Staining

Prepare Gram stains of each unknown culture. Record your observations. Can you identify these organisms from the Gram stain? _____
Can you differentiate these organisms based on the Gram stain? _____

F.A. Test

1. Prepare a smear of each culture on clean glass slides and allow the smears to air dry.
2. Fix the smears by immersing in 95% ethyl alcohol for 2 minutes.
3. Drain off excess alcohol and quickly rinse the slides in F.A. buffer.
4. Apply 2 drops of fluorescent antibody to each smear and spread the antibody by tilting the slide.
5. Place the slides in a Petri plate containing moistened filter paper for 30 minutes. The moistened filter paper will prevent evaporation during incubation.
6. Remove the slides and air dry.
7. Place a drop of F.A. mounting medium on each smear and set a cover slip on the mounting medium.
8. Examine each slide using a brightfield microscope and record your observations.
9. Examine each slide using a fluorescent microscope and look for fluorescence. (See color plate IX.2.) Record your observations.

EXERCISE 44

Fluorescent-Antibody Technique

Name _____

Date _____

Lab Section _____

Purpose _____

_____

_____

Data

Gram stain results:

Culture # _____

Culture # _____

F.A. test results.

Culture	Microscopic Appearance	
	Brightfield	Fluorescent
# _____		
# _____		

Questions

1. Which culture was *E. coli*? _____

2. Briefly outline the procedure you would have to follow to identify the *E. coli* culture without fluorescent

   microscopy. _____

   _____

   _____

   _____

   _____

3. Was the test performed in this exercise a direct or an indirect F.A. test? _____

4. Compare and contrast direct and indirect F.A. tests. _____

_____

_____

_____

_____

5. Outline a procedure using fluorescent-antibody testing for identifying the other culture used in this exercise.

_____

_____

PART 12

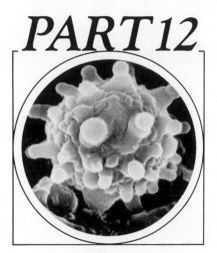

Microorganisms and Disease

Many microorganisms grow abundantly both inside and on the surface of the normal adult body. The microorganisms that establish more or less permanent residence without producing diseases are known as **normal flora** (see the figure). Microorganisms that may be present for a few days or months are called **transient flora.**

The relationship between normal flora and a healthy person may be classified as one of two types of **symbiosis** (the close association between two organisms): **commensalism** or **mutualism.** In a *commensal* relationship, one organism is benefited and the other is unharmed. For example, many of the bacteria living in the large intestine are supplied with continual food and a constant temperature, while the host is neither benefited nor harmed. In a *mutualistic* relationship, both organisms are benefited. Some bacteria that live on and in our bodies receive a constant supply of nutrients, while they, in exchange, supply something of benefit to us. For example, *E. coli* synthesizes vitamin K and certain B vitamins for its human host.

A third kind of symbiosis is **parasitism.** In this relationship, one organism gains while the other is harmed. In a disease state, the microorganism is a parasite and is harming its human host. An organism like this is called a **pathogen.**

The pathogens that cause disease consist of many different organisms. Robert Koch speculated in one of his early publications:

> *On this . . . I take my stand, and, till the cultivation of bacteria from spore to spore shows that I am wrong, I shall look on pathogenic bacteria as consisting of different species . . . .**

Some members of the normal flora become parasites under certain conditions. These organisms are called **opportunists.**

In order to cause a disease, a pathogen must gain access to the human body. The route of access is called the **portal of entry.** Some microbes cause disease only when they gain access by the "correct" portal of entry.

In Exercises 45 through 49, we will examine the variety of bacteria associated with the human body. In a clinical laboratory, samples from diseased tissue are cultured, and pathogens must be distinguished from normal and transient flora. In Exercise 50, we will identify bacteria in a simulated clinical sample.

*Quoted in T. D. Brock, ed. *Milestones in Microbiology.* Washington, DC: American Society of Microbiology, 1961, p. 99.

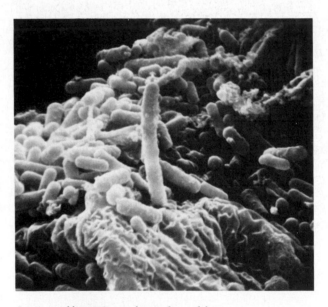

Lactic acid bacteria on the surface of the stomach epithelium (3900 ×).

EXERCISE 45

Bacteria of the Skin

Objectives

After completing this exercise you should be able to:

1. Isolate and identify bacteria from the human skin.
2. Provide an example of normal skin flora.
3. List characteristics used to identify the staphylococci.
4. Explain why many bacteria are unable to grow on human skin.

Background

The skin is generally an inhospitable environment for most microorganisms. The dry layers of keratin-containing cells that make up the epidermis (the outermost layer of the skin) are not easily colonized by most microbes. Sebum, secreted by oil glands, inhibits bacterial growth, and salts in perspiration create a hypertonic environment. Perspiration and sebum are nutritive for certain microorganisms, however, which establishes them as part of the normal flora of the skin.

Normal flora of the skin vary according to skin region. The flora of the face reflect that of the throat just behind the mouth, and the flora of the anal region are influenced by microorganisms of the lower gastrointestinal tract. More bacteria are found in moist areas, such as the axilla (armpit) and the sides of the nose, than on the dry surfaces of arms or legs. Transient flora are present on hands and arms in contact with the environment.

Propionibacterium live in hair follicles on sebum from oil glands. The propionic acid they produce maintains the pH of the skin between 3 and 5, which suppresses the growth of other bacteria. Most bacteria on the skin are gram-positive. Catalase-positive, salt-tolerant members of gram-positive cocci (*Bergey's Manual*, Gram-Positive Cocci, family Micrococcaceae) are the most frequently encountered bacteria on the skin.

Staphylococcus aureus is part of the normal flora of the skin and is also considered a pathogen. Strains producing **coagulase,** an enzyme that coagulates (clots) the fibrin in blood, are pathogenic. A test for the presence of coagulase is used to distinguish *S. aureus* from other species of *Staphylococcus*.

Although many different bacterial genera live on human skin, we will attempt to isolate and identify a member of the family Micrococcaceae (catalase-positive) in this exercise.

Materials

Petri plates containing mannitol salt agar (2)

Sterile cotton swab

Sterile saline

3% hydrogen peroxide

Gram staining reagents

Fermentation tubes, coagulase plasma (as needed)

Techniques Required

Exercises 5, 11, 14, 17, and 21.

Procedure

1. Wet the swab with saline, and push against the wall of the test tube to express excess saline (Figure 45.1*a*). Swab any surface of your skin. The sides of the nose, axilla, an elbow, or a pus-filled sore are possible areas.
2. Swab one-half of the plate with the swab. Using a sterile loop, streak back and forth into the swabbed area a few times, then streak away from the inoculum (Figure 45.1*b*).

> **☣ Discard the swab in the disinfectant.**

3. Incubate the plate inverted at 35°C for 24 to 48 hours.
4. Examine the colonies. Members of the Micrococcaceae usually form large opaque colonies. Record the appearance of the colonies and any mannitol fermentation (yellow halos). (See color plate V.2.) Gram stain the colonies and test for catalase production (Exercise 17). Perform the catalase test by making a suspension of the desired colony on a slide and adding a drop of H_2O_2.
5. Subculture a catalase-positive, gram-positive coccus on another mannitol salt plate. Do not attempt

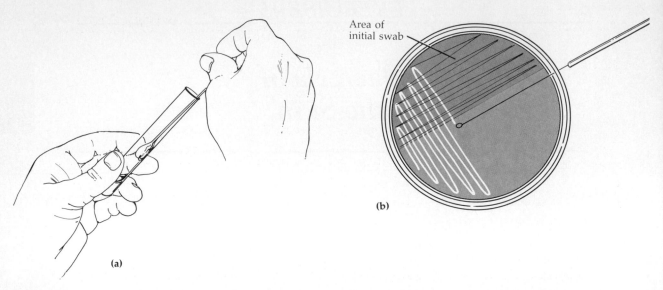

Figure 45.1 _____

Taking a sample from the skin. **(a)** Aseptically moisten a sterile cotton swab in saline. **(b)** Using a sterile loop, spread the inoculum over the remaining half of the plate.

Table 45.1

Key to the Micrococcaceae

Gram-positive cocci	
↓	
Catalase-positive	
↓	
I. Cells arranged in tetrads	*Micrococcus*
A. Glucose not fermented	
1. Colonies have yellow pigment	*M. luteus*
2. Colonies have red pigment	*M. roseus*
3. Colonies have yellow-red pigment	*Planococcus*
II. Cells arranged in grapelike clusters, glucose fermented	*Staphylococcus*
A. Acid produced from mannitol, coagulase-positive	*S. aureus*
B. No acid from mannitol, coagulase-negative	
1. Acid produced from trehalose	*S. saprophyticus*
2. No acid from trehalose	*S. epidermidis*

to subculture a colony that has not been tested for catalase. Why?_____

6. Using the key to the Micrococcaceae in Table 45.1, proceed to identify your isolate.

   a.   To test for coagulase, place a loopful of rehydrated coagulase plasma on a clean slide. Add a loopful of water and make a heavy suspension of the bacteria to be tested. Observe for clumping of the bacterial cells (clumping = coagulase-positive; no clumping = coagulase-negative). (See color plate VIII.2.)

   b.   Inoculate the appropriate fermentation tubes.

EXERCISE 45

Bacteria of the Skin

Name _____

Date _____

Lab Section _____

Purpose _____

_____

_____

Data

Source of inoculum: _____

First Mannitol Salt Plate

	Colony		
	1	2	3
Colony Description			
Pigment			
Mannitol fermentation			

Colony description

Which colony was isolated? _____

  Catalase reaction: _____

  Gram stain

    Reaction: _____

    Cell morphology: _____

    Cell arrangement: _____

Additional Biochemical Tests

_____

_____

_____

Conclusion

What organism did you identify? _____

Questions

1. Why is mannitol salt agar used as a selective medium for normal skin flora? _____

    _____

    _____

2. After you have observed a gram-positive coccus, what is the additional information you need before performing

    a coagulase test? _____

    _____

3. List three identifying characteristics of *Staphylococcus aureus*. _____

    _____

    _____

4. List three factors that protect the skin from infection. _____

    _____

    _____

    _____

5. What is coagulase? How is it related to pathogenicity? _____

    _____

    _____

    _____

Bacteria of the Respiratory Tract

Objectives

After completing this exercise you should be able to:

1. List representative normal flora of the respiratory tract.
2. Differentiate the pathogenic streptococci based on biochemical testing.
3. List a characteristic used to identify *Neisseria, Corynebacterium, Mycobacterium,* and *Bordetella.*

Background

The respiratory tract can be divided into two systems: the upper and lower respiratory systems. The **upper respiratory system** consists of the nose and throat, and the **lower respiratory system** consists of the larynx, trachea, bronchial tubes, and alveoli. The lower respiratory tract is normally sterile because of the efficient functioning of the ciliary escalator. The upper respiratory system is in contact with the air we breathe — air contaminated with microorganisms.

The throat is a moist, warm environment, allowing many bacteria to establish residence. Species of many different genera — such as *Staphylococcus, Streptococcus, Neisseria,* and *Haemophilus* — can be found living as normal flora in the throat. Despite the presence of potentially pathogenic bacteria in the upper respiratory system, the rate of infection is minimized by microbial **antagonism.** Certain microorganisms of the normal flora suppress the growth of other microorganisms through competition for nutrients and production of inhibitory substances.

Streptococcal species are the predominant organisms in throat cultures, and some species are the major cause of bacterial sore throats (acute pharyngitis). Streptococci are identified by biochemical characteristics, including hemolytic reactions and antigenic characteristics (Lancefield's system). Hemolytic reactions are based on hemolysins that are produced by streptococci while growing on blood-enriched agar. Blood agar is usually made with defibrinated sheep blood (5.0%), sodium chloride (0.5%) to minimize spontaneous hemolysis, and nutrient agar. Three patterns of hemolysis can occur on blood agar:

1. **β-hemolysis:** Complete hemolysis, giving a clear zone with a clean edge around the colony. (See color plate VIII.4.)

2. **α-hemolysis:** Incomplete hemolysis, producing methemoglobin and a green, cloudy zone around the colony. (See color plate VIII.3.)
3. **γ-hemolysis:** No hemolysis, and no change in the blood agar around the colony.

Streptococci that are α-hemolytic and γ-hemolytic are usually normal flora, whereas β-hemolytic streptococci are frequently pathogens.

The streptococci can be antigenically classified into Lancefield groups A through O by antigens in their cell walls. Over 90% of streptococcal infections are caused by β-hemolytic Group A streptococci. *S. pyogenes,* a β-hemolytic group A *Streptococcus,* is sensitive to the antibiotic Bacitracin; other streptococci are resistant to Bacitracin. Optochin sensitivity is used to distinguish *S. pneumoniae* from other α-hemolytic streptococci.

We will examine other bacteria found in the throat in the third part of this exercise.

Materials

Throat Culture
Petri plate containing blood agar

Sterile cotton swab

Gram staining reagents

Hydrogen peroxide, 3%

Streptococcus
Petri plate containing blood agar

Sterile cotton swabs (2)

Forceps and alcohol

Optochin disk

Bacitracin disk

10% bile salts

Tubes containing 2 ml nutrient broth (2)

Selected Characteristics of Other Bacteria of the Respiratory Tract
Petri plate containing tellurite agar

Petri plate containing brain-heart infusion agar

Gram staining reagents

Methylene blue (simple stain)

Acid-fast staining reagents

Oxidase reagent

CO_2 jar

Cultures

Streptococcus

Streptococcus pyogenes

Streptococcus pneumoniae

Selected Characteristics of Other Bacteria of the Respiratory Tract

Corynebacterium diphtheriae (avirulent)

Mycobacterium smegmatis on Lowenstein Jensen medium

Bordetella bronchiseptica on Bordet–Gengou medium

Neisseria sicca

Techniques Required

Exercises 3, 5, 6, 11, 17, and 25.

Procedure

Throat Culture

> **☣ Work only with swabs collected from your own throat.**

1. Swab your throat with a sterile cotton swab. The area to be swabbed is between the "golden arches" (glossopalatine arches), as shown in Figure 46.1. Do not hit the tongue.
2. After obtaining an inoculum from the throat, swab one-half of a blood agar plate. Streak the remainder of the plate with a sterile loop. (See Figure 45.1*b*.)

> **☣ Discard the swab in disinfectant.**

3. Incubate the plate inverted at 35°C for 24 hours. Observe the plate for hemolysis. Transfer some colonies to a slide and perform a catalase test. Why can't the catalase test be done on blood agar? ____
____

Streptococcus

1. Inoculate each half of a blood agar plate, one side with *S. pyogenes* and the other half with *S. pneumoniae*. Use a swab for confluent growth.

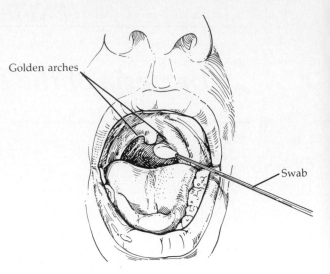

Figure 46.1 ____
Swab your throat with a sterile swab.

> **☣ Be careful — these bacteria are pathogens.**

2. Dip forceps in alcohol and burn off the alcohol.

> **⚠ Keep the forceps tip pointed down while burning. Keep the beaker of alcohol away from the flame.**

Using the forceps, place and press a Bacitracin disk and an Optochin disk on each half. Space the disks so that zones of inhibition (Exercise 25) may be observed.
3. Incubate the plates inverted at 35°C for 24 hours. Observe for hemolysis and inhibition of growth by Bacitracin and Optochin.
4. Prepare Gram stains from smears of each organism. Do the organisms differ microscopically? ____
5. Using a sterile loop, prepare a suspension of each organism in a tube of nutrient broth.
6. Add a few drops of 10% bile salts to each tube. Observe the tubes after 15 minutes for lysis of the cells. Which culture is "bile soluble"? ____

Selected Characteristics of Other Bacteria of the Respiratory Tract

1. Perform a Gram stain on smears of each organism, and an acid-fast stain (Exercise 6) on *Mycobacterium* and one other organism.
2. Record the appearance of *Mycobacterium* on Lowenstein Jensen medium.
3. Prepare a simple stain of *Bordetella*. Do the cells stain darker in one region? ____

4. Inoculate *Corynebacterium* onto tellurite agar and incubate inverted at 35°C for 24 hours. Record any pigment production. Do a simple stain of *Corynebacterium*. Describe the shape of the bacterium.

5. Inoculate brain-heart infusion agar with *Neisseria*. Incubate the plate inverted in a CO_2 jar at 35°C for 24 hours. Prepare a CO_2 jar (Figure 46.2) by placing the plates to be incubated with a candle into a large jar. Light the candle and screw the lid on the jar. The candle will extinguish when the CO_2 concentration in the jar reaches 5% to 10%.

6. Perform an oxidase test on *Neisseria* (Exercise 17). To test for the production of cytochrome oxidase, follow step a or step b.

   a. Flood an agar culture with oxidase reagent (1% tetramethyl-*p*-phenylene-diamine). Observe for a change in the color of the *colonies*, not the agar. The presence of oxidase is indicated by pink to black colonies.

   b. Moisten an oxidase disk and place it *on* colonies on a plate. Incubate the plate for 15 minutes at 35°C. The disk and colonies turn black in oxidase-positive cultures. (See color plate III.9.)

Figure 46.2 —————————————————
A CO_2 jar. Light the candle and screw the lid onto the jar. Do not let the flame touch the Petri plates.

EXERCISE 46

Bacteria of the Respiratory Tract

Name _____

Date _____

Lab Section _____

Purpose _____

_____

_____

Data

Throat Culture

	Colony			
	1	2	3	4
Appearance of colonies on blood agar				
Hemolysis				
Catalase reaction				

Streptococcus

	Organism	
	S. pyogenes	*S. pneumoniae*
Gram stain Gram reaction		
Morphology		
Arrangement		
Blood agar plate Appearance of colonies		
Hemolysis		

Continued

	Organism	
	S. pyogenes	*S. pneumoniae*
Inhibition by Optochin		
Bacitracin		
Bile solubility		

Other Bacteria

Characteristics	*Mycobacterium smegmatis*	*Bordetella bronchiseptica*	*Cornyebacterium diphtheriae*	*Neisseria sicca*
Gram stain Gram reaction				
Morphology				
Arrangement				
Acid-fast reaction				
Appearance of cells in simple stain	na			na
Oxidase reaction	na	na	na	
Appearance of colonies on Lowenstein Jensen medium		na	na	na
Bordet-Gengou medium	na		na	na
Tellurite agar	na	na		na
Brain-heart infusion agar	na	na	na	

na = not applicable

Questions

1. Why is blood agar a differential medium? _____

_____

2. If a Bacitracin disk is added to the area swabbed on a blood agar plate, a rapid identification technique (Exercise 18) results. Why? _____

_____

_____

3. Is the Gram stain of significant importance in the identification of the organisms studied in this exercise? _____ Explain. _____

_____

4. You have isolated a gram-positive coccus from a throat culture that you cannot identify as staphylococci or streptococci. A test for one enzyme can be used to distinguish *quickly* between these bacteria. What is the enzyme? _____

5. Name one disease caused by each genus used in this exercise. Identify the species that causes that disease.

_____

_____

_____

_____

_____

_____

6. Design a rapid identification technique for *S. pneumoniae.* _____

_____

_____

_____

_____

EXERCISE 47

Bacteria of the Mouth

EXERCISE
47

Objectives

After completing this exercise you should be able to:

1. List characteristics of streptococci found in the mouth.
2. Describe the formation of dental caries.
3. Explain the relationship between sucrose and caries.

Background

The mouth contains millions of bacteria in each milliliter of saliva. Some of these bacteria are transient flora carried on food. Some species of *Streptococcus* are part of the normal flora of the mouth. An identification scheme for streptococci found in the mouth is shown in Table 47.1.

Streptococcus mutans, S. salivarius, and *S. sanguis* produce sticky polysaccharides specifically from su-

crose. Bacterial exoenzymes hydrolyze sucrose into its component monosaccharides, glucose and fructose. The energy released in the hydrolysis is used by *S. mutans* and *S. sanguis* for polymerization of the glucose to form **dextran.** Fructose released from the sucrose is fermented to produce lactic acid. Enzymes of *S. salivarius* have similar specificity for sucrose, but the fructose is polymerized into **levan,** and the liberated glucose is fermented.

The dextran capsule enables the bacteria to adhere to surfaces in the mouth. *S. salivarius* colonizes the surface of the tongue, and *S. sanguis* and *S. mutans,* the teeth. Masses of bacterial cells, dextran, and debris adhering to the teeth constitute dental plaque. Production of lactic acid by bacteria in the plaque initiates dental caries by eroding tooth enamel. Streptococci and other bacteria, such as *Lactobacillus,* are able to grow on the exposed dentin and tooth pulp.

Other carbohydrates, such as glucose or starch, may

Table 47.1
Key to Streptococci Found in the Mouth

Gram-positive cocci in chains
↓
Catalase-negative
↓
No growth at 10°C, in 6.5% NaCl broth, or at pH 9.6
↓
α- or γ-hemolysis

I. Inulin fermented

  A. Raffinose fermented

    1. Mannitol fermented   *S. mutans*
    2. Mannitol not fermented   *S. salivarius*

  B. Raffinose not fermented

    1. Acetoin produced (V–P test)   *S. sobrinus*
    2. Acetoin not produced (V–P test)   *S. sanguis*

II. Inulin not fermented

  A. Mannitol fermented

    1. Esculin hydrolyzed   *S. milleri*
    2. Esculin not hydrolyzed   *S. oralis*

  B. Mannitol not fermented

    1. Esculin hydrolyzed   *S. mitis*
    2. Esculin not hydrolyzed   *S. mitior*

be fermented by bacteria, but they are not converted to dextran and hence do not promote plaque formation. "Sugarless" candies contain mannitol or sorbitol, which cannot be converted to dextran, although they may be fermented by such normal flora as *S. mutans*. Acid production increases the size of dental caries.

The number of *S. mutans* present in paraffin-stimulated saliva has been correlated with the potential for the formation of caries. In this exercise, we will observe and count *S. mutans* and other polysaccharide-producing streptococci in a paraffin-stimulated saliva sample. The media used contain sucrose to promote capsule formation, and Mitis-Salivarius-Bacitracin (MSB) agar inhibits the growth of most oral bacteria, except *S. mutans*.

Materials

Rodac plate containing MSB agar

Petri plate containing sucrose blood agar

Paraffin

Tongue blade

50-ml beaker

1-ml pipettes (3)

99-ml saline dilution blanks (2)

Hockey stick

Beaker containing alcohol

CO_2 jar

3% hydrogen peroxide

Techniques Required

Exercises 10 and 17; and Appendices A and E.

Procedure*

> ☣ **Work only with your own saliva, and discard all contaminated materials in disinfectant.**

Getting Ready

1. Label an MSB plate and a sucrose blood agar plate.
2. Allow a small piece of paraffin to soften under your tongue, then chew it for 1 minute. Do not swallow your saliva or the paraffin.

*Adapted from C. Hoover. "Indigenous Flora of Saliva and Supragingival Plaque." Unpublished exercise for dental students. San Francisco: University of California, Department of Stomatology, n.d.

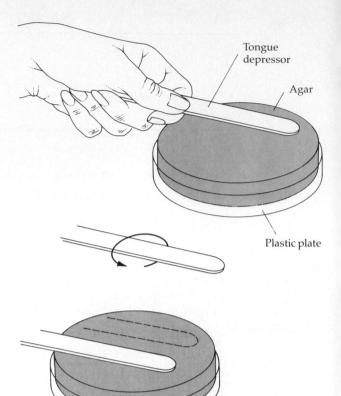

Figure 47.1
To inoculate the Rodac plate, press one side of the tongue blade on the agar. Turn the blade over, and press the other side on the remaining half of the agar.

MSB Agar

1. Rotate a sterile tongue blade in your mouth ten times so both sides of the blade are thoroughly inoculated with saliva. Remove the tongue blade through closed lips to remove excess saliva. Don't swallow yet.
2. Press one side of the tongue blade onto one-half of the surface of the MSB agar. Then, press the other side of the blade onto the other side of the agar, as shown in Figure 47.1.
3. Incubate the plate inverted in a CO_2 jar (see Figure 46.2) for 72 to 96 hours at 35°C.
4. Examine the MSB plate with a dissecting microscope (Appendix E) for the presence of *S. mutans*. *S. mutans* colonies are light blue to black, raised, and rough; their surface resembles etched glass or "burnt" sugar. Other bacteria occasionally grow on MSB; therefore, careful observation of colony morphology is necessary to accurately estimate the number of *S. mutans* present.

Sucrose Blood Agar

1. Collect your saliva in a 50-ml beaker. Don't be embarrassed; everyone makes about 2 liters of saliva over the course of a day!
2. Label two dilution blanks "1" and "2."
3. Pipette 1 ml of saliva into dilution blank 1. Cap the bottle and shake it. This will give a dilution of 1: _____ . Using another pipette, transfer 1 ml of dilution 1 into dilution blank 2 and invert several times to mix. This will give a total dilution of 1: _____ .
4. With another pipette, transfer 0.1 ml from dilution 2 to the surface of a sucrose blood agar plate.
5. Disinfect a hockey stick by dipping in alcohol, quickly igniting the alcohol in a Bunsen burner flame, and letting the alcohol burn off.

> ⚠ **After spreading, dip the hockey stick in alcohol and burn off the alcohol. Keep the flame away from the beaker of alcohol.**

6. Using the cooled hockey stick, spread the fluid over the entire surface of the agar (see Figure 27.2).
7. Disinfect the hockey stick and return it.
8. Incubate the plate inverted in a CO_2 jar (see Figure 46.2) for 72 to 96 hours at 35°C.
9. Examine the sucrose blood agar plate for production of large raised mucoid (gum-drop) colonies; these are likely to be *S. salivarius*. Look carefully for glistening colonies surrounded by an indentation of the agar surface that cannot be moved without tearing the agar; these are likely to be *S. sanguis*. *S. mutans* colonies, if present, may be recognized by the presence of a drop of liquid polysaccharide on top of or surrounding the colony. Determine whether catalase (Exercise 17) is produced by suspected streptococci. The catalase test cannot be done directly on agar containing blood. Why not? _____

_____

_____

EXERCISE 47

Bacteria of the Mouth

Name _____

Date _____

Lab Section _____

Purpose _____
_____
_____

Data

MSB Agar

Species	Number	Description

Sucrose Blood Agar

	Colony		
	1	2	3
Appearance of colonies			
Catalase production			
Species			

Conclusions

Compare your data with those of your classmates and draw a conclusion about your oral flora. _____
_____
_____
_____

Questions

1. List three characteristics of streptococci found in the mouth. _____
_____

_____

2. *S. salivarius* forms large mucoid colonies on sucrose blood agar. Would you expect the same type of colonies on

glucose blood agar? _____ Briefly explain your answer. _____

_____

3. Studies have shown that both sucrose and bacteria are necessary for tooth decay. Why should this be true? _____
_____

_____

_____

4. Is MSB agar selective or differential? _____ Explain. _____

_____

5. Is sucrose blood agar selective or differential? _____ Explain. _____

_____

6. Even if a large number of *S. mutans* is in your saliva, how might you avoid tooth decay? _____

_____

_____

_____

Bacteria of the Gastrointestinal Tract

Objectives

After completing this exercise you should be able to:

1. List the bacteria commonly found in the gastrointestinal tract.
2. Define the following terms: coliform, enteric, and enterococci.
3. Interpret results from TSI slants and SF broth.

Background

The stomach and small intestine have relatively few microorganisms, as a result of the hydrochloric acid produced by the stomach and the rapid movement of food through the small intestine. But microbial populations in the large intestine are enormous, exceeding 10^{11} bacteria per gram of feces. Most of these intestinal organisms are commensals, and some are in mutualistic relationships with their human hosts. Some intestinal bacteria synthesize useful vitamins, such as folic acid and vitamin K. The normal intestinal flora prevent colonization of pathogenic species by producing antimicrobial substances and competition.

The population of the large intestine consists primarily of anaerobes of the genera *Bacteriodes, Bifidobacterium, Lactobacillus,* and such facultative anaerobes as *Escherichia, Enterobacter, Citrobacter,* and *Proteus.* Catalase-negative, gram-positive cocci of the genus *Enterococcus* are also present. (These species were once classified in the genus *Streptococcus.*)

Most diseases of the gastrointestinal system result from the ingestion of food or water that contains pathogenic microorganisms. Good sanitation practices, modern methods of sewage treatment, and the disinfection of drinking water help break the fecal–oral cycle of disease. A number of tests have been developed to identify facultative anaerobes associated with fecal contamination (see Exercises 18, 51, and 52). Gram-negative, facultatively anaerobic rods are a large and diverse group of bacteria that includes the **enteric family** (Enterobacteriaceae).

Media have been developed to differentiate between lactose-fermenting enterics and nonlactose-fermenting enterics. The lactose fermenters are called **coliforms** (Exercise 51) and are generally not pathogenic. The nonlactose-fermenting group includes such pathogens

as *Salmonella* and *Shigella*. One of the most common media is eosin-methylene blue (EMB) agar. EMB agar is selective in that the eosin-methylene blue dyes are inhibitory to gram-positive organisms; thus, they allow the medium to selectively culture gram-negative organisms. EMB is differential, in that lactose-fermenting organisms (coliforms) give colored colonies and nonlactose fermenters produce colorless colonies.

After isolation on EMB agar, differential screening media such as **triple sugar iron (TSI)** agar can be used to further categorize organisms. TSI contains:

0.1% glucose

1.0% lactose

1.0% sucrose

0.02% ferrous sulfate

Phenol red

Nutrient agar

As shown in Figure 48.1, if the organism ferments only glucose, the tube will turn yellow in a few hours. The bacteria quickly exhaust the limited supply of glucose and start oxidizing amino acids for energy, giving off ammonia as an end-product. Oxidation of amino acids increases the pH, and the indicator in the slanted portion of the tube will turn back to red. The butt will remain yellow. If the organism in the TSI slant ferments lactose and/or sucrose, the butt and slant will turn yellow and remain yellow for days due to the increased level of acid production. Gas production by an organism can be ascertained by the appearance of bubbles *in* the agar. TSI can also be used to indicate whether hydrogen sulfide (H_2S) (Exercise 16) has been produced due to the reduction of sulfur-containing compounds. H_2S reacts with ferrous sulfate in the medium, producing ferrous sulfide, a black precipitate. Table 48.1 shows how enteric bacteria can be identified using EMB, TSI, and additional tests.

The presence of enterococci can be used to indicate fecal contamination. Enterococci are fairly specific for their mammalian hosts. *Enterococcus faecalis* (formerly called *Streptococcus faecalis*) is found in humans, *S. bovis* in cattle, and *S. equinus* in horses. **SF (*Streptococcus faecalis*) broth** can be used to de-

Red

Red

(a) No sugar fermentation

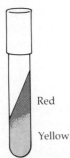

Red

Yellow

Gas

Red

Yellow

Without gas production

With gas production

(b) Glucose fermented; lactose and sucrose not fermented

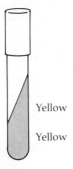

Yellow

Yellow

Gas

Yellow

Yellow

Yellow

Without gas production

With gas production

(c) Glucose and lactose and/or sucrose fermented

Black precipitate

(d) H₂S production can occur in addition to **(a)**, **(b)**, and **(c)**

Figure 48.1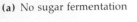

Reactions in triple sugar iron (TSI) agar after 24 hours' incubation.

tect the presence of enterococci. SF broth contains sodium azide to inhibit the growth of gram-negative bacteria. Enterococci will grow and ferment the glucose in the broth. Growth and a change in the bromcresol purple indicator from purple (pH 6.6) to yellow (pH 5.2) may indicate the presence of enterococci.

Tomato juice agar is an enriched medium used to encourage the growth of lactobacilli. *Lactobacillus* does not grow well on ordinary media, but its growth is enhanced by the addition of tomato juice and peptonized milk.

Materials

Petri plate containing EMB agar

Petri plate containing tomato juice agar

Tube containing SF broth

TSI slant

Sterile water

Brewer anaerobic jar and GasPak®

Tube containing sterile cotton swab

10-ml pipette

Fecal sample (animal or human feces; swab feces or the anus after a bowel movement with a sterile swab and bring the contaminated swab to class in a sterile tube)

Gram stain reagents

Techniques Required

Exercises 5, 11, 14, 17, and 19.

Procedure

1. If the feces are hard, break apart in a sterile Petri plate and emulsify in sterile water. Swab a small area on the EMB with an inoculated swab, then streak for isolation with a loop. (See Figure 45.1*b*.)
2. Incubate the plate inverted at 35°C for 24 to 48 hours.
3. Using the fecal swab, inoculate tomato juice agar. Incubate the plate inverted anaerobically (Exercise 19) at 35°C for 24 to 48 hours.

> **Discard the test tube, tube of water, and other contaminated materials in disinfectant.**

Table 48.1

Identification Scheme for Enteric Genera Primarily Using EMB and TSI

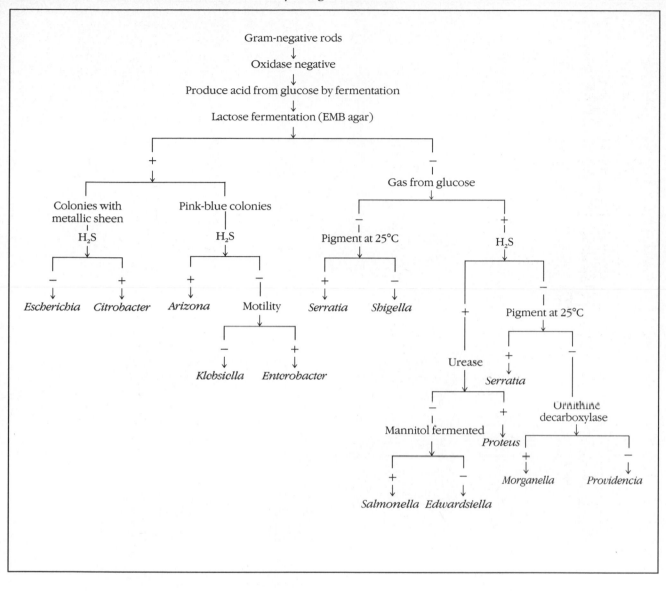

4. Inoculate SF broth by placing the fecal swab in the broth. Incubate the SF broth at 35°C for 24 to 48 hours.

5. Observe your EMB plate and those of other students. *Escherichia* and *Citrobacter* produce a green metallic sheen by reducing the dye. Did you see it? _____ Record the appearance and colors of the colonies. (See color plates V.4, V.5, and V.6.) Prepare a Gram stain from an isolated colony.

6. Select one isolated colony using an inoculating needle and inoculate the TSI slant. Streak the sur-

face of the slant, then *stab* into the butt of the agar. Incubate the tube at 35°C for 24 hours. Observe and record results from TSI slants inoculated from different appearing colonies. (See color plate III.10.)

7. Observe the tomato juice agar for the presence of *Lactobacillus*. Perform a Gram stain and a catalase test (Exercise 7) on isolated colonies. Is *Lactobacillus* catalase-positive or catalase-negative? _____

8. Observe the SF broth for the presence of enterococci. Gram stain.

EXERCISE 48

Bacteria of the Gastrointestinal Tract

Name _____

Date _____

Lab Section _____

Purpose _____

_____

_____

Data

EMB Agar

Source of inoculum: _____

	Colony			
	1	2	3	4
Appearance of colonies				
Lactose fermentation?				

TSI

Colony picked: _____

Appearance: Slant _____

              Butt _____

              Gas? _____

              H_2S _____

What reactions occurred in the TSI? _____

_____

_____

_____

Other TSI slants: _____

_____

Description of colonies picked: _____

Results of TSI slant:

Tomato Juice Agar

Description of colonies: _____

_____

Gram reaction of colony picked: _____

Catalase test: _____

SF Broth

Growth _____ Acid _____

Gram stain: _____

Questions

1. What genus do you think you have in your TSI slant? _____

What additional information do you need to identify it? _____

_____

2. Did you culture *Lactobacillus* on the tomato juice agar? _____

How do you know? _____

3. What can you conclude from the SF broth? _____

_____

4. What tests, if any, would you need to perform to identify bacteria producing a green metallic sheen on EMB

agar? _____

5. Differentiate between a coliform, an enteric that is not a coliform, and an enterococcus. Give an example of an

organism in each group. _____

_____

_____

6. Is EMB a selective medium? _____ A differential medium? _____

Bacteria of the Urogenital Tract

Objectives

After completing this exercise you should be able to:

1. List bacteria found in urine from a healthy individual.
2. Identify, through biochemical testing, bacteria commonly associated with urinary tract infections.
3. Determine the presence or absence of *Neisseria gonorrhoeae* in a G.C. smear.

Background

The urinary and genital systems are closely related anatomically, and some diseases that affect one system also affect the other system, especially in the female. The upper urinary tract and urinary bladder are normally sterile. The urethra does contain resident bacteria including *Streptococcus, Bacteroides, Mycobacterium, Neisseria,* and enterics. Most bacteria in urine are the result of contamination by skin flora during passage. The presence of bacteria in urine is not considered to be a urinary tract disease unless there are at least 10^5 bacteria per milliliter of urine.

Many infections of the urinary tract, such as cystitis (inflammation of the urinary bladder) or pyelonephritis (inflammation of the kidney), are caused by opportunistic pathogens and are related to fecal contamination of the urethra, and to medical procedures, such as catheterization.

Standard examination of urine consists of a plate count on blood agar for total number of organisms, coupled with a streak plate of undiluted urine on EMB agar (Exercise 48). Why?_____

_____

A "clean-catch" collection of a voided urine specimen depends on careful cleaning of the external urogenital surface with an antiseptic solution and collection of a midstream specimen in a sterile container.

In the first part of this exercise, you will examine normal urine. In the second part, three gram-negative rods that commonly cause cystitis will be provided to you. *E. coli* and *Proteus* are enterics. *E. coli* is one of the coliforms (Exercise 51). *Proteus* is actively motile and exhibits "swarming" on solid media (Exercise 7), where the cells at the periphery move away from the main colony. *Pseudomonas* is a gram-negative aerobic rod. *Pseudomonas aeruginosa* is commonly found in the soil and other environments. Under the right conditions, particularly in weakened hosts, this organism

can cause urinary tract infections, burn and wound infections, and abscesses. *P. aeruginosa* infections are characterized by blue-green pus. This bacterium produces an extracellular, water-soluble pigment called **pyocyanin** ("blue pus") that diffuses into its growth medium.

Most diseases of the genital system are transmitted by sexual activity and are therefore called **sexually transmitted diseases (STDs).** Most of the bacterial diseases can be readily cured with antibiotics if treated early and can largely be prevented by the use of condoms. Nevertheless, STDs are a major U.S. public health problem.

The most common reportable communicable disease in the United States is gonorrhea, an STD caused by the gram-negative diplococci *Neisseria gonorrhoeae* called gonococci, or G.C. Gonorrhea is diagnosed by identifying the organism in the pus-filled discharges of patients, as demonstrated with G.C. smears in the third part of this exercise. Cultures from patients' discharges can be made on Thayer–Martin medium and incubated in a CO_2 jar (see Figure 46.2). An oxidase test (Exercise 17) is performed on characteristic colonies for confirmation.

Materials

Urine Culture
Petri plate containing blood agar

Petri plate containing EMB agar

Sterile wide-mouth jar, approximately 50 to 250 ml

Sterile 1-ml pipettes (3)

Sterile 0.9-ml saline dilution blanks (2)

Hockey stick and alcohol

Cystitis
Petri plate containing EMB agar

Petri plate containing Pseudomonas agar P

Tubes containing OF-glucose medium (4)

Urea agar slants (2)

Mineral oil

Oxidase reagent

Gram stain reagents

Cultures

Tube containing *Escherichia coli, Proteus vulgaris,* and *Pseudomonas aeruginosa*

G.C. Smears

G.C. smears

One unknown smear # _____

Techniques Required

Exercises 5, 10, 11, 13, 15, 17, and 48; and Appendices A and B.

Procedure

Urine Culture

> ☣ **Work only with urine collected from your own body.**

1. Collect a "clean-catch" urine specimen, as described by your instructor, using the sterile jars available in the lab. Refrigerate until ready to perform step 2. Why?_____

   _____

2. Add 0.1 ml of urine to one dilution blank, and pipette up and down three times to mix. With a new pipette, transfer 0.1 ml to the second tube and mix. Transfer 0.1 ml to the surface of the blood agar plate. This is a 1: _____ dilution.

3. Dip the spreading rod (hockey stick) in alcohol, and flame off the excess alcohol to disinfect. Spread the 0.1 ml of diluted urine on the blood agar with the hockey stick. (See Figure 28.2.)

> ⚠ **After spreading, dip the hockey stick in alcohol and ignite. Keep the flame away from the beaker of alcohol.**

4. Streak a loopful of *undiluted* urine on EMB agar.

> ☣ **Discard the jar and pipettes in disinfectant.**

5. Incubate both plates inverted at 35°C for 24 to 48 hours.

6. Count the colonies on the blood agar plate, and determine the number of bacteria per milliliter of urine:

$$\text{Bacteria/ml} = \frac{\text{Number of colonies}}{\text{Amount plated} \times \text{dilution}}$$

7. Examine the EMB plate for the presence of possible coliforms. Consult your instructor if you are alarmed by your results.

Cystitis

1. Inoculate an EMB plate and Pseudomonas agar P plate with the mixed culture of bacteria.

2. Incubate the plates inverted at 35°C for 24 to 48 hours. (See color plates V.3, V.4, V.5, and V.6.)

3. Examine the EMB plate for lactose-fermenting colonies. Which of the three organisms ferments lactose? _____
   Look for swarming. Which organism is actively motile on solid media?_____
   Are any small, nonlactose-fermenting colonies present? _____

4. Prepare a Gram stain from each different colony type.

5. Examine the Pseudomonas agar P plate. Can you identify *Pseudomonas?* _____

6. Perform on oxidase test on each different organism (see Exercise 17).

7. Inoculate four tubes of OF-glucose medium: two with *Proteus* and two with *Pseudomonas.* Plug one tube of each organism with mineral oil (Exercise 13). Inoculate a urea agar slant with each organism (Exercise 15). Incubate the tubes at 35°C for 24 to 48 hours.

8. Record the results of the OF-glucose and urease production tests. Why isn't it necessary to perform these two biochemical tests on *E. coli?*_____

   _____

G.C. Smears

1. Examine the G.C. smears provided.
2. Determine whether *N. gonorrhoeae* could be present in your unknown G.C. smear.

EXERCISE 49

Bacteria of the Urogenital Tract

Name _____

Date _____

Lab Section _____

Purpose _____

_____

_____

Data

Urine Culture

Blood agar plate:

  Number of colonies: _____

  Total count (bacteria/ml): _____

  Hemolysis: _____

EMB plate:

  Possible coliforms present: _____

  If so, describe the colonies: _____

  Noncoliforms: _____

Cystitis

	Organisms		
	E. coli	*P. vulgaris*	*P. aeruginosa*
EMB agar			
Appearance of colonies	_____	_____	_____
Lactose fermentation	_____	_____	_____
Swarming	_____	_____	_____
Gram stain	_____	_____	_____
Pseudomonas agar P	_____	_____	_____
Appearance of colonies	_____	_____	_____
Pigmentation	_____	_____	_____
Oxidase reaction	_____	_____	_____
OF-glucose (Fermentation or oxidation)	_____	_____	_____
Urease production	_____	_____	_____

G.C. Smears

Diagram the appearance of a known G.C.-positive smear.   Diagram the appearance of a known G.C.-negative smear.

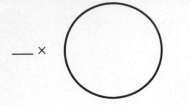

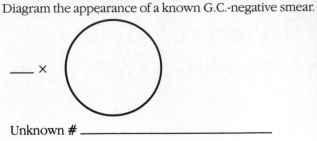

Unknown # _____

Diagram the appearance of your unknown smear.

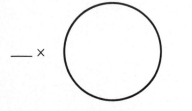

Could *N. gonorrhoeae* be present? _____

Questions

1. Why is EMB agar inoculated with a urine specimen? _____

_____

2. With the results obtained on the blood agar and EMB plates, is a urinary tract disease possible? _____

_____

3. Differentiate between the pigment of *P. aeruginosa* on Pseudomonas agar P and the "pigment" of *E. coli* on EMB

   agar without referring to the colors. _____

_____

4. The enterics and pseudomonads look alike microscopically. How can you easily distinguish between these two

   groups of bacteria? _____

_____

_____

5. Why are females more prone to urinary tract infections than males? _____

_____

6. What role does antibiotic treatment have in yeast infections of the urinary tract? _____

_____

Identification of an Unknown from a Clinical Sample

Unknown #3*

*So simple did it seem to me
The day t'was first presented,
A Sherlock Holmes of lab technique
No effort be resented.*

*Now six Gram stains, three cultures later
The first step barely taken
Gram vari rods, both red and blue
My confidence was shaken.*

*Straight rods 'tis true, yet red and blue
In ones and twos and chain
Somes spores would show the truth of it
But stains were all in vain.*

*Alas! No spores of vivid green
Would show in fields of red.
Ah Ha! Too soon! Spores take some time
Flashed through my seeking head.*

*With rod in hand, though young it be,
It surely must, said I,
Reveal some trait, another test
Than just a non-cocci.*

*So while I wait don't waste the time
The sugars show some trait.
So off to these, ferment or not
I can't just sit and wait.*

*Glucose a plus, Sucrose the same
No gas, and lactose zilch.
The spores again with no green spots
It's on to litmus milk.*

*So mannitol and nitrate too
I'll run, for these should show,
Though spores resist at now day six,
Some little cells should grow.*

*Day seven's gone and spores still hide
From vision sharply tooled.
But agents stained from tests complete
Have many others ruled.*

*Divorce now near, by child disowned
On job, the thought "to fire"
I'll take my stand and flunk or not
B. subtilis, or tests are liars.*

MARY DAVIS

Objectives

After completing this exercise you should be able to:

1. Separate and identify two bacteria from a sample.
2. Determine the unknown bacteria's sensitivity to chemotherapeutic agents.

Background

Clinical samples often contain many microorganisms, and a pathogen must be separated from resident normal flora. Differential and selective media are used to facilitate isolation of the pathogen, after which it must be identified. Also, antibiotic sensitivity tests are performed to aid the physician in prescribing treatment. Since effective treatment depends on identification of the pathogen, a clinical laboratory must provide results

as quickly and as accurately as possible. Microbiology laboratory results are usually available within 48 hours.

In this exercise, you will be provided with a simulated clinical sample containing two bacteria. The procedure will differ from that used in a clinical laboratory in that you will be asked to isolate and identify both organisms in the sample. After you have separated the two bacteria in pure culture, prepare stock and working cultures. To obtain pure cultures, inoculate a trypticase soy agar plate *and* a differential medium. Select the differential medium using the type of sample as a clue. For instance, a fecal sample should probably be inoculated onto _____ agar. Since differential media should give the information needed to isolate the organisms, why is a trypticase soy agar inoculated also? _____
_____

Materials

Petri plates containing trypticase soy agar

*M. E. Davis. *ASM News* 42(3):164, 1976. Reprinted with the permission of the American Society for Microbiology.

Table 50-1
Zone-Diameter Interpretive Standards

Antimicrobial Agent	Disk Content	Zone Diameter (mm)	
		Resistant	Susceptible
Ampicillin	10 μg		
Gram-negative enterics and enterococci		11 or less	14 or more
Staphylococci		20 or less	29 or more
Chloramphenicol	30 μg	12 or less	18 or more
Erythromycin	15 μg	13 or less	18 or more
Penicillin-G	10 U		
Staphylococci		20 or less	29 or more
Other bacteria		11 or less	22 or more
Streptomycin	10, 250, or 300 μg	11 or less	15 or more
Sulfathiazole	250 μg	10 or less	16 or more
Sulfisoxazole	150 μg	12 or less	17 or more
Sulfonamides	300 μg	12 or less	17 or more
Tetracycline	30 μg	14 or less	19 or more

Source: Adapted from A. L. Barry. "Zone-Diameter Interpretive Standards and Approximate MIC Correlates." *The Antimicrobic Susceptibility Test: Principles and Practices.* Ann Arbor, MI: University Microfilm International, n.d., Table 14-2.

Trypticase soy agar slants

Mueller–Hinton agar

All stains, reagents, and media previously used

Antimicrobial disks

Culture

Unknown sample # _____

Techniques Required

Exercises 25, 32, and 45–49.

Procedure

Getting Started

1. Refer to Exercise 32 for general information regarding identification of an unknown. Study the identification schemes in Exercises 32, 45, 47, and 48 as you proceed.
2. You may be able to prepare a Gram stain directly from the unknown if it does not contain a high concentration of organic matter to interfere with the stain results.
3. Streak the unknown onto a trypticase soy agar plate and an appropriate differential medium. *Do not be*

wasteful. Inoculate only the necessary media. Incubate the plates at 35°C for 24 to 48 hours.

Second Period

1. Examine the plates and select colonies that differ from each other in appearance and that are separated for easy isolation. (See color plate X.)
2. Streak each organism for isolation onto half of a trypticase soy agar plate. Incubate the plate at 35°C for 24 to 48 hours. This plate can be your first working culture. Prepare a stock culture of each organism.

Third Period

1. Prepare a Gram stain of each organism. What should you do if you do not think you have a pure culture? _____
   _____
2. After determining the staining and morphologic characteristics of the two organisms, determine which biochemical tests you will need. Plan your work carefully. The same tests need not be performed on both bacteria.
3. Record which tests were performed and the results.
4. Use the Mueller–Hinton agar and antimicrobial disks to determine whether your unknowns are resistant (R) or sensitive (S) to the antimicrobial agents listed in Table 50.1. (See color plate VIII.1.)

EXERCISE 50

Identification of an Unknown from a Clinical Sample

Name _____

Date _____

Lab Section _____

Purpose _____

_____

_____

Data

Unknown # _____

Source: _____

Preliminary Gram stain results: _____

	Organism	
	A	B
Trypticase soy agar plate		
Appearance of colonies	_____	_____
Differential media		
(_____)		
Appearance of colonies	_____	_____
Additional information	_____	_____
Gram stain		
Morphology	_____	_____
Gram reaction	_____	_____
Trypticase soy agar slant		
Appearance of colonies	_____	_____
Other tests:		
_____	_____	_____
_____	_____	_____
_____	_____	_____
_____	_____	_____

| | Organism | |
	A	B
Antibiotic sensitivity to		
Ampicillin	_____	_____
Chloramphenicol	_____	_____
Erythromycin	_____	_____
Penicillin G	_____	_____
Streptomycin	_____	_____
Sulfonamide	_____	_____
Tetracycline	_____	_____

Questions

1. What two bacterial species were in unknown # _____ ? _____

_____

2. On a separate sheet of paper, write your rationale for arriving at the conclusion.

3. Why would a clinical microbiology laboratory perform antimicrobial susceptibility tests as well as identify the

unknown? _____

_____

PART 13

Microbiology and the Environment

―――――――――――― *EXERCISES* ――――――――――――

It should be evident from previous exercises that microorganisms are omnipresent in our environment. However, the presence of some microorganisms in certain places, such as food or water, may be undesirable. In Exercises 51 and 52, we will use the presence of nonpathogenic bacteria to indicate water pollution. The presence and number of bacteria in foods can indicate that foods were contaminated during processing; these bacteria may result in food spoilage (Exercise 53).

Harmful microorganisms are only a very small fraction of the total microbial population. The activities of most microbes are in fact beneficial. In Exercise 54, we will use selected microorganisms to produce desired flavors in foods and to preserve foods.

The activities of soil bacteria are essential to the maintenance of life on Earth (see the figure). Nitrogen fixation (Exercise 55) was first described in 1893 by

Sergei Winogradsky, who entered into the study of biogeochemical cycles because he was

> *impressed by the incomparable glitter of Pasteur's discoveries. I started to investigate the great problem of fixation of atmospheric nitrogen. I succeeded without too much difficulty in isolating an anaerobic rod called* Clostridium *that would perform this function.**

Winogradsky named the organism *Clostridium pasteurianum* in honor of Pasteur. Some microbes are able to degrade dangerous pollutants in the environment (Exercise 56).

*Quoted in H. A. Lechevalier and M. Solotorovsky. *Three Centuries of Microbiology.* New York: Dover Publications, 1974, p. 263.

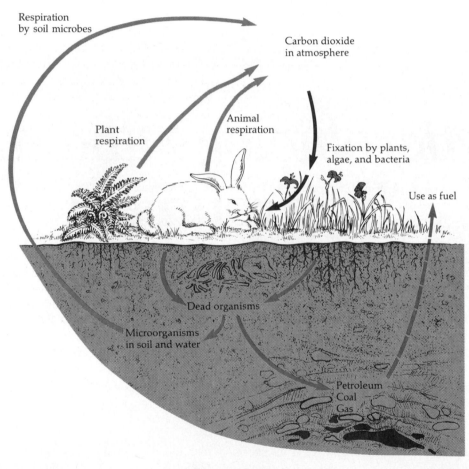

The carbon cycle.

Microbes in Water: Multiple-Tube Technique

Objectives

After completing this exercise you should be able to:

1. Define coliform.
2. Provide the rationale for determining the presence of coliforms.
3. List and explain each step in the multiple-tube technique.

Background

Tests that determine the bacteriologic quality of water have been developed to prevent transmission of waterborne diseases of fecal origin. However, it is not practical to look for pathogens in water supplies, because pathogens occur in such small numbers that they might be missed by sampling. Moreover, when pathogens are detected, it is usually too late to prevent occurrence of the disease. Rather, the presence of **indicator organisms** is used to detect fecal contamination of water. An indicator organism must be present in human feces in large numbers and must be easy to detect. The most frequently used indicator organisms are the coliform bacteria. **Coliforms** are aerobic or facultatively anaerobic, gram-negative, nonendospore-forming, rod-shaped bacteria that ferment lactose with acid and gas formation within 48 hours at 35°C. Coliforms are not usually pathogenic, although they can cause opportunistic infections (Exercise 49). Coliforms are not restricted to the human gastrointestinal tract but may be found in other animals and in the soil. Tests that determine the presence of fecal coliforms (of human origin) have been developed.

Established public health standards specify the maximum number of coliforms allowable in each 100 ml of water, depending on the intended use of the water (for example, drinking, water-contact sports, or treated waste water for irrigation or for discharge into a bay or river).

Coliforms can be detected and enumerated in the **multiple-tube technique** (Figure 51.1). In this method, coliforms are detected in three stages. In the **presumptive test,** dilutions from a water sample are added to lactose fermentation tubes. The lactose broth can be made selective for gram-negative bacteria by the addition of lauryl sulfate or brilliant green and bile.

Fermentation of lactose to acid and gas is a positive reaction.

Samples from the positive presumptive tube at the highest dilution are streaked onto EMB (eosin-methylene blue) agar (Exercise 48) in the **confirmed test. EMB agar** is selective because the EMB dyes inhibit the growth of gram-positive organisms, allowing the growth of gram-negative bacteria. EMB is differential in that lactose-fermenting bacteria give colored colonies and nonlactose fermenters produce colorless colonies. Colored colonies on EMB agar is a positive confirmed test.

In the **completed test,** isolated lactose-positive colonies from EMB agar are inoculated into lactose broth and onto a nutrient agar slant. If acid and gas are produced in the lactose broth, and the isolated bacterium is a gram-negative nonendospore-forming rod, it is a positive completed test.

The number of coliforms is determined by a statistical estimation called the **most probable number (MPN)** method. In the presumptive test, tubes of lactose broth are inoculated with samples of the water being tested. A count of the number of tubes showing acid and gas is then taken, and the figure compared to statistical tables, shown in Table 51.1. The number is the *most probable number* of coliforms per 100 ml of water.

We will use a multiple-tube technique in this exercise. In Exercise 52, we will use a membrane filter technique to determine if coliforms are present.

Materials

9-ml single-strength lactose fermentation tubes (7)

20-ml 1.5-strength lactose fermentation tubes (3)

Petri plate containing EMB agar

Nutrient agar slant

Sterile 10-ml pipette

Sterile 1-ml pipette

Gram staining reagents

Water sample, 50 ml (Bring your own, from a pond or stream.)

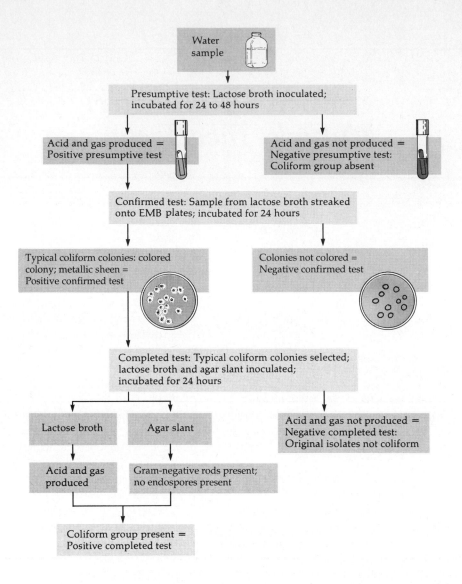

Figure 51.1 _____

Analysis of drinking water for coliforms by the multiple-tube technique.

Techniques Required

Exercises 5, 10, 11, 14, and 48; Appendix A.

Procedure

1. Label three single-strength lactose broth tubes "0.1," label another three tubes "1," and label the three 1.5-strength broth tubes "10."
2. Inoculate each 0.1 tube with 0.1 ml of your water sample.
3. Inoculate each 1 tube with 1.0 ml of your water sample.
4. Inoculate each 10 tube with 10 ml of the water sample. Why is 1.5-strength lactose broth used for this step? _____
   _____
5. Incubate the tubes for 24 to 48 hours at 35°C.

6. Record the results of your presumptive test (Figure 51.1). Which tube has the highest dilution of the water sample? _____
   _____

   Determine the number of coliforms per 100 ml of the original sample using Table 51.1. If a tube has gas, streak the EMB agar with the positive broth. Incubate the plate inverted for 24 to 48 hours at 35°C.

7. Record the results of your confirmed test (Figure 51.1). (See color plates V.4, V.5, and V.6.) How can you tell whether coliform colonies are present? __
   _____

   If coliform colonies are present, inoculate a single-strength lactose broth tube and nutrient agar slant from an isolated colony. Incubate the broth and slant for 24 to 48 hours at 35°C, and check for the presence of coliforms.

8. Gram stain the organisms found on the slant.

Table 51.1

Most Probable Numbers (MPN) Index for Various Combinations of Positive and Negative Results
When Three 10-ml Portions, Three 1-ml Portions, and Three 0.1-ml Portions Are Used

Number of Tubes Giving Positive Reaction out of			MPN Index per 100 ml	Number of Tubes Giving Positive Reaction out of			MPN Index per 100 ml
3 of 10 ml Each	3 of 1 ml Each	3 of 0.1 ml Each		3 of 10 ml Each	3 of 1 ml Each	3 of 0.1 ml Each	
0	0	0	<3	3	0	0	23
0	0	1	3	3	0	1	39
0	1	0	3	3	0	2	64
1	0	0	4	3	1	0	43
1	0	1	7	3	1	1	75
1	1	0	7	3	1	2	120
1	1	1	11	3	2	0	93
1	2	0	11	3	2	1	150
2	0	0	9	3	2	2	210
2	0	1	14	3	3	0	240
2	1	0	15	3	3	1	460
2	1	1	20	3	3	2	1,100
2	2	0	21	3	3	3	≥2,400
2	2	1	28				

Source: Standard Methods for the Examination of Water and Wastewater, 13th ed. New York:
American Public Health Association, 1971.

EXERCISE 51

Microbes in Water: Multiple-Tube Technique

Name _____

Date _____

Lab Section _____

Purpose _____

_____

_____

Data

Water sample source: _____

Presumptive Test

Inoculum	Number of Tubes With:			Possible Coliforms Present?
	Growth	Acid	Gas	
0.1 ml				
1.0 ml				
10 ml				

MPN: _____

Confirming Test

Tube: _____ Growth: _____

Appearance of colonies: _____

_____

Are coliforms suspected? _____

Completed Test

Appearance of colony used: _____

Gas from lactose: _____

Gram reaction and cell morphology: _____

_____

_____

Data from water samples tested by other students:

Sample	MPN

Questions

1. Were coliforms present in your water sample? _____

2. What is the MPN of your water sample? _____ per _____ ml

3. Why isn't a pH indicator needed in the lactose broth fermentation tubes? _____
    _____

4. Why are coliforms used as indicator organisms if they are not usually pathogens? _____
    _____

5. Could the water have a high concentration of the pathogenic bacterium *Salmonella* and give negative results in
    the multiple-tube technique? Briefly explain. _____
    _____

6. Why don't we inoculate EMB agar directly and bypass lactose broth? _____
    _____
    _____

Microbes in Water: Membrane Filter Technique

Objectives

After completing this exercise you should be able to:

1. Explain the principle of the membrane filter technique.
2. Perform a coliform count using the membrane filter technique.

Background

In Exercise 51, we observed that fecal contamination of water can be determined by the number of coliforms present in a water sample, by the multiple-tube technique. Coliforms can also be detected by the membrane filter technique.

In the **membrane filter technique,** water is drawn through a thin filter (see Appendix F). Filters with a variety of pore sizes are available. Pores of 0.45 μm are used for filtering out most bacteria. Bacteria are retained on the filter, which is then placed on a pad of suitable nutrient medium. Nutrients that diffuse through the filter can be metabolized by bacteria trapped on the filter. Each bacterium that is trapped on the filter will develop into a colony. Bacterial colonies growing on the medium can be counted. When a selective or differential medium (Exercise 12) is used, desired colonies will have a distinctive appearance. Endo medium is frequently used as a selective and differential medium with the membrane filter technique. The composition of Endo medium is shown in Table 52.1. The desoxycholate and lauryl sulfate select for gram-negative bacteria. Coliforms ferment lactose, causing their colonies to have a red color or metallic sheen from the fuchsin sulfite indicator. Nonfermenting bacteria grow on the polypeptides and produce colorless colonies.

Materials

47-mm Petri plate containing EMB or Endo agar

Sterile membrane filter apparatus

Sterile 0.45 μm filter

Forceps

Alcohol

Table 52.1

Chemical Composition of Bacto-*m* Endo Broth MF®

Yeast extract	0.15%
Casitone	0.5%
Thiopeptone	0.5%
Tryptose	1.0%
Lactose	1.25%
Sodium desoxycholate	0.01%
Dipotassium phosphate	0.4375%
Monopotassium phosphate	0.1375%
Sodium chloride	0.5%
Sodium lauryl sulfate	0.005%
Sodium sulfite	0.21%
Basic fuchsin	0.105%

Source: Difco Laboratories. Prepared according to the Millipore Corporation.

Sterile pipette or graduated cylinder, as needed

Sterile rinse water

Water sample (Bring your own, from a pond or stream.)

Techniques Required

Exercise 10; Appendices A and F.

Procedure

1. Set up filtration equipment (see Appendix F). Remove wrappers as each piece is fitted into place. Why shouldn't all the wrappers be removed at once? _____

    a. Attach the filter trap to the vacuum source. What is the purpose of the filter trap? _____

    b. Place the filter holder base (with stopper) on the filtering flask. Attach the flask to the filter trap.

> ⚠️ **Disinfect the forceps by burning off the alcohol. Keep the beaker of alcohol away from the flame.**

   c. Using the sterile forceps, place a filter on the filter holder. Why must the filter be centered exactly on the filter holder? _____

_____

   d. Set the funnel on the filter holder and fasten in place.

2. Filtering

   a. Shake the water sample and pour or pipette a measured volume into the funnel. Your instructor will help you determine the volume. *(For samples of 10 ml or less, pour 20 ml sterile water into the funnel first.)*

   b. Turn on the vacuum and allow the sample to pass into the filtering flask. Leave the vacuum on.

   c. Pour sterile rinse water into the funnel. *(Use the same volume as the sample.)* Allow the rinse water to go through the filter. Turn the vacuum off.

3. Inoculation (Figure 52.1)

   a. Carefully remove the filter from the filter holder using sterile forceps. Why does the filter have to be "peeled" off? _____

_____

   b. Carefully place the filter on the culture medium. Do not bend the filter; place one edge down first, then carefully set the remainder down. Place the filter on the plate as it was in the filter holder.

4. Invert the plate and incubate for 24 hours at 35°C.

5. Examine the plates for the presence of coliforms. On Endo or EMB, coliforms will form pink to red colonies, and some may have a green metallic sheen. (See color plate V.7.) Count the number of coliform colonies:

Number of coliforms per 100 ml of water =

$$100 \times \frac{\text{Number of coliform colonies}}{\text{Volume of sample filtered}}$$

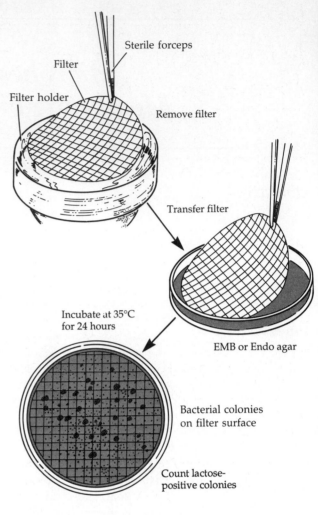

Figure 52.1 ————————————————

Inoculation. Using sterile forceps, remove the filter from the filter holder. Place the filter on the culture medium, gradually laying it down from one edge to the other.

EXERCISE 52

Microbes in Water: Membrane Filter Technique

Name _____

Date _____

Lab Section _____

Purpose _____

_____

_____

Data

Sample tested: _____

Describe the general appearance of the colonies on _____ medium: _____

_____

_____

_____

Number of coliform colonies: _____

Number of coliforms/100 ml water: _____
Use the space below for your calculations:

Data from water samples tested by other students:

Sample	Coliforms/100 ml

Questions

1. Which water sample(s) is (are) potable? _____

2. Which water sample(s) is (are) contaminated with fecal material? _____

3. Can you determine the source of contamination (for example, human, domestic animal, wild animal) from this test? _____

4. What basic assumption is made in this technique, if the number of bacteria is determined from the number of colonies? _____

   _____

5. Why can filtration be used to sterilize culture media and other liquids? _____

   _____

6. If you did Exercise 51, list one advantage and one disadvantage of each method of detecting coliforms. _____

   _____

   _____

   _____

   _____

   _____

7. Why is the membrane filter technique useful for a sanitarian working in the field? _____

   _____

8. Design an experiment to detect the presence and number of *Acetobacter* in wine. Why would you want to perform such a test? _____

   _____

   _____

Microbes in Food: Contamination

Objectives

After completing this exercise you should be able to:

1. Determine the number of bacteria in a food sample using a standard plate count.
2. Provide reasons for monitoring the bacteriologic quality of foods.
3. Explain why the standard plate count is used in food quality control.

Background

Illness and food spoilage can result from microbial growth in foods. The sanitary control of food quality is concerned with testing foods for the presence of pathogens. During processing (grinding, washing, and packaging), food may be contaminated with soil microbes and flora from animals, food handlers, and machinery.

Foods are the primary vehicle responsible for the transmission of diseases of the digestive system. For this reason, they are examined for the presence of coliforms. Recall that the presence of coliforms indicates fecal contamination (Exercise 51).

Standard plate counts are routinely performed on food and milk by food processing companies and public health agencies. The **standard plate count** is used to determine the total number of viable bacteria in a food sample. The presence of large numbers of bacteria is undesirable in most foods because it increases the likelihood that pathogens will be present, and it increases the potential for food spoilage.

In a standard plate count, the number of **colony-forming units (c.f.u.)** is determined. Each colony may arise from a group of cells rather than from one individual cell. The initial sample is diluted through serial dilutions (Appendix B) in order to obtain a small number of colonies on each plate. A known volume of the diluted sample is placed in a sterile Petri plate and melted, and cooled nutrient agar is poured over the inoculum. After incubation, the number of colonies is counted. Plates with between 25 and 250 colonies are suitable for counting. A plate with fewer than 25 colonies is inaccurate because a single contaminant could influence the results. A plate with more than 250 colonies is extremely difficult to count. The microbial pop-

ulation in the original food sample can then be calculated using the following equation:

Colony-forming units/gram or ml of sample =

$$\frac{\text{Number of colonies}}{\text{Amount plated} \times \text{dilution*}}$$

A limitation of the standard plate count is that only bacteria capable of growing in the culture medium and environmental conditions provided will be counted. A medium that supports the growth of most heterotrophic bacteria is used.

Materials

Melted standard plate count or nutrient agar, cooled to 45°C

Sterile 1-ml pipettes (Part A, 2; Part B, 3)

Sterile Petri plates (Part A, 4; Part B, 4)

Sterile 99-ml dilution blanks (Part A, 1; Part B, 2)

Food samples, diluted 1:10

Techniques Required

Exercise 11; Appendices A and B.

Procedure

First Period. A: Bacteriologic Examination of Milk

1. Obtain a sample of either raw or pasteurized milk that has been diluted 1:10.
2. Using a sterile 1-ml pipette, aseptically transfer 1 ml of the 1:10 milk sample into a 99-ml dilution blank; label the bottle "$1:10^3$" and discard the pi-

*"Dilution" refers to the tube prepared by serial dilutions (Appendix B). For example, if 250 colonies were present on the $1:10^7$ plate, the calculation would be as follows:

$$\text{Colony-forming units/gram} = \frac{250 \text{ colonies}}{0.1 \text{ ml} \times 10^{-6}}$$
$$= 250 \times 10^6 \times 10^1$$
$$= 2,500,000,000$$
$$= 2.5 \times 10^9$$

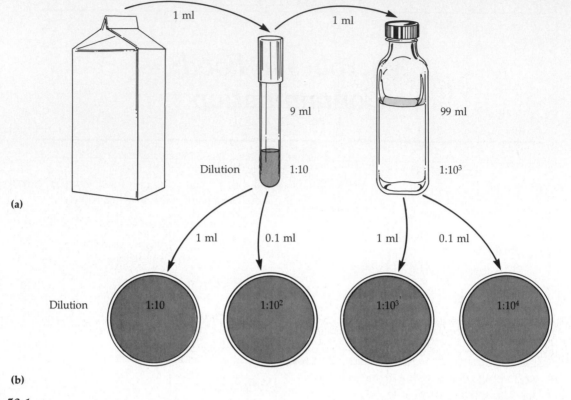

(a)

(b)

Figure 53.1

Standard plate count of milk. **(a)** Make serial dilutions of a milk sample. **(b)** Mark four Petri plates with the dilutions.

pette (Figure 53.1*a*). Shake the bottle twenty times, with your elbow resting on the table, as shown in Figure 53.2.

3. Label the bottoms of four sterile Petri plates with the dilutions: "1:10," "1:10²," "1:10³," and "1:10⁴" (Figure 53.1*b*).

4. Using a 1-ml pipette, aseptically transfer 0.1 ml of the 1:10³ dilution into the bottom of the 1:10⁴ plate. *Note:* 0.1 ml of the 1:10³ dilution results in a 1:10⁴ dilution of the original sample. Using the same pipette, transfer 1.0 ml of the 1:10³ into the plate labeled 1:10³. Pipette 0.1 ml and 1.0 ml from the 1:10 dilution into the 1:10² and 1:10 plates, respectively, with the same pipette (Figure 53.3*a*). Why is it important to proceed from the highest to the lowest dilution? _____

5. Check the temperature of the water bath containing the nutrient agar. Test the temperature of the agar container with your hand. It should be "baby bottle" warm. Why? _____

6. Pour the melted nutrient agar into one of the plates (to about one-third full) (Figure 53.3*b*). Cover the plate and swirl it gently (Figure 53.3*c*) to distribute the milk sample evenly through the agar. Continue until all the plates are poured.

7. When each plate has solidified, invert it, and incubate all plates at 35°C for 24 to 48 hours.

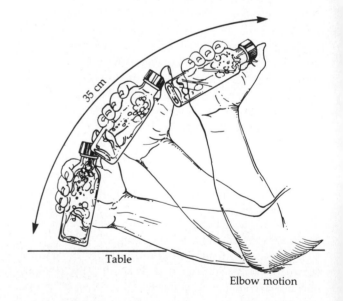

Table

Elbow motion

Figure 53.2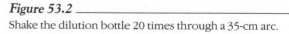

Shake the dilution bottle 20 times through a 35-cm arc.

First Period. B: Bacteriologic Examination of Hamburger and Frozen Vegetables

1. Obtain a sample of raw hamburger or frozen vegetables diluted 1:10.

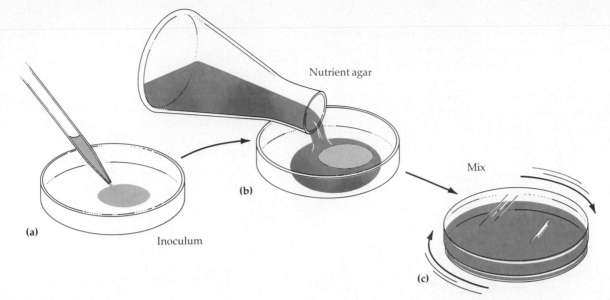

Figure 53.3

Pour plate. **(a)** Pipette inoculum into a Petri plate. **(b)** Add liquefied nutrient agar, and **(c)** mix the agar with the inoculum by gentle swirling of the plate.

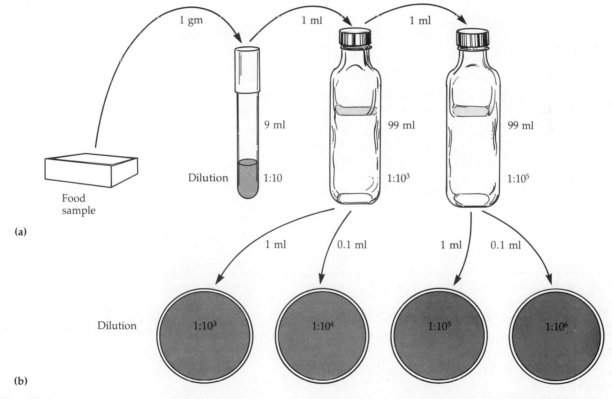

Figure 53.4

Standard plate count on food. **(a)** Prepare serial dilutions of a food sample. **(b)** Label four Petri plates with the dilutions to be plated.

2. Using a sterile 1-ml pipette, aseptically transfer 1 ml of the 1:10 sample into a 99-ml dilution blank; label the bottle "1:10³" and discard the pipette. Shake the bottle twenty times, with your elbow resting on the table, as shown in Figure 53.2. Make a 1:10⁵ dilution using another 99-ml dilution blank (Figure 53.4*a*). Shake as before.

3. Label the bottoms of four sterile Petri plates with

the dilutions: "1:10^3," "1:10^4," "1:10^5," and "1:10^6" (Figure 53.4*b*).

4. Using a 1-ml pipette, aseptically transfer 0.1 ml of the 1:10^5 dilution into the 1:10^6 plate. *Note:* 0.1 ml of a 1:10^5 dilution results in a 1:10^6 dilution of the original sample. Using the same pipette, repeat this procedure with the 1:10^3 dilution until all the plates have been inoculated (Figure 53.3*a*). Why can the same pipette be used for each transfer? _____

    _____

5. Check the temperature of the water bath containing the nutrient agar. Test the temperature of the agar container with your hand. It should be "baby bottle" warm. Why? _____

    _____

6. Pour the melted nutrient agar into one of the plates (to about one-third full) (Figure 53.3*b*). Cover the plate and swirl it gently (Figure 53.3*c*) to distribute the sample through the agar evenly. Continue until all the plates are poured.

7. When each plate has solidified, invert it, and incubate all plates at 35°C for 24 to 48 hours.

Second Period

1. Arrange each plate in order from lowest to highest dilution.

2. Select the plate with 25 to 250 colonies. Record data for plates with fewer than 25 colonies as *too few to count (TFTC)* and those with more than 250 colonies, *too many to count (TMTC)*.

3. Count the number of colonies on the plate selected.

4. Multiply the number of colonies by the dilution of that plate to determine the number of bacteria in the original food. For example, if 129 colonies were counted on a 1:10^3 dilution:

$$\frac{129 \text{ colonies}}{1 \text{ ml} \times 10^{-3}} = 129,000$$

$$= 1.29 \times 10^5 \text{ colony-forming units/} \\ \text{ml or gram of food}$$

EXERCISE 53

Microbes in Food: Contamination

Name _____

Date _____

Lab Section _____

Purpose _____

_____

_____

Data

Sample _____

Dilution	Colonies per Plate
1·	
1:	
1:	
1:	

Number of colony-forming units per ml (or gram) of original food: _____
Do your calculations in the space below.

Record data for other foods tested by other students.

Food	Bacteria/ml (or gram)

Questions

1. What could you do to ensure that the bacteria present in foods do not pose a health hazard? _____

   _____

   _____

2. Are bacteria identified in the standard plate count? _____

   Why is a standard plate count performed on food? _____

   _____

   _____

3. Why are plates with 25 to 250 colonies used for calculations? _____

   _____

   _____

4. In a quality-control laboratory, each dilution is plated in duplicate or triplicate. Why would this increase the

   accuracy of a standard plate count? _____

   _____

   _____

5. There are other techniques for counting bacteria, such as a direct microscopic count and turbidity. Why is the

   standard plate count preferred for food? _____

   _____

   _____

6. Why is ground beef a better bacterial growth medium than a steak or roast? _____

   _____

7. List four foodborne diseases. Identify one that is transmitted primarily via milk. _____

   _____

   _____

   _____

8. Why does repeated freezing and thawing increase bacterial growth in meat? _____

   _____

   _____

Microbes Used in the Production of Foods

Objectives

After completing this exercise you should be able to:

1. Explain how the activities of microorganisms are used to preserve food.
2. Define fermentation.
3. Produce an enjoyable product.

Background

Microbial fermentations are used to produce a wide variety of foods. **Fermentation** means different things to different people. In industrial usage, it is any large-scale microbial process occurring with or without air. To a biochemist, it is the group of metabolic processes that release energy from a sugar or other organic molecule, do not require oxygen or an electron transport system, and use an organic molecule as a final electron acceptor. In this exercise, we will examine a lactic acid fermentation used in the production of food.

In dairy fermentations, such as yogurt production, microorganisms use lactose and produce lactic acid without using oxygen. In nondairy fermentations, such as wine production, yeast use sucrose to produce ethyl alcohol and carbon dioxide under anaerobic conditions. If oxygen is available, the yeast will grow aerobically, liberating carbon dioxide and water as metabolic end-products.

Historically, milk has been selectively fermented, with the resulting acidity preventing spoilage by acid-intolerant microbes. These "sour" milks have varied from country to country depending on the source of milk, conditions of culture, and microbial "starter" used. Milk from donkeys to zebras has been used, with the Russian *kumiss* (horse milk), containing 2% alcohol, and Swedish *surmjölk* (reindeer milk) being unusual examples. The bacteria yield lactic acid, and the yeast produce ethyl alcohol. Currently, two fermented cow milk products, buttermilk and yogurt, are widely used.

Buttermilk is the fluid left after cream is churned into butter. Today, buttermilk is actually prepared by souring true buttermilk or by adding bacteria to skim milk and then flavoring it with butterflake. *Streptococcus lactis* ferments the milk, producing lactic acid (sour); and neutral fermentation products (diacetyls) are produced by *Leuconostoc*. Yogurt originated in the Balkan countries, goat milk being the primary source. Yogurt is milk that has been concentrated by heating and then fermented at elevated temperatures. *Streptococcus* produces lactic acid, and *Lactobacillus* produces the flavors and aroma of yogurt.

Techniques Required

Exercises 3, 5, 10, and 11; Appendix A.

Materials

Homogenized milk

Nonfat dry milk

Large beaker

Stirring rod

Thermometer

Glass test tube

Hot plate or ring stand and asbestos pad

5-ml pipette

Styrofoam or paper cups with lids

Plastic spoons

Petri plate containing trypticase soy agar

pH paper

Gram stain reagents

Optional: jam, jelly, honey, and so on

Culture

Commercial yogurt or *Streptococcus thermophilus* and *Lactobacillus bulgaricus*

Procedure

Be sure all glassware is clean.

1. Add 100 ml of milk per person (in your group) to a wet beaker (wash out the beaker first with water to decrease the sticking of the milk.) Put a thermometer in a glass test tube containing water before placing in the beaker.

2. Heat the milk on a hot plate or over a burner on an asbestos pad placed on a ring stand, to about 80°C for 10 to 20 minutes. Stir occasionally. Do not let it boil. Why is the milk heated? _____

   _____

3. Cool to about 65°C and add 3 g of nonfat dry milk per person. Stir to dissolve. Why is dry milk added? _____

   _____

4. Rapidly cool to about 45°C. Pour milk equally into the cups.

5. Inoculate each cup with 1 to 2 teaspoonfuls of commercial yogurt, or 2.5 ml *S. thermophilus* and 2.5 ml *L. bulgaricus*. Cover and label.

6. Incubate at 45°C for 4 to 18 hours or until firm (custardlike).

7. Cool the yogurt to about 5°C. Save a small amount for steps 8 through 10. Then taste with a clean spoon. Add jam or some other flavor if you desire. Eat!

☣ Do not work with other bacteria or perform other exercises while eating the yogurt.

---

8. Determine the pH of the yogurt.

9. Make a smear, and after heat fixing, Gram stain it. Record your results.

10. Streak for isolation on nutrient agar. Incubate the plate inverted at 45°C.

11. After distinct colonies are visible, record your observations. Prepare Gram stains from each different colony.

EXERCISE 54

Microbes Used in the Production of Foods

Name _____

Date _____

Lab Section _____

Purpose

_____

_____

_____

Observations

Yogurt characteristics:

Taste: _____

Consistency: _____

Odor: _____

pH: _____

Gram stain results: _____

_____

Streak plate results: _____

_____

_____

Gram stain of isolated colonies: _____

_____

Questions

1. How can pathogens enter yogurt, and how can this be prevented? _____

   _____

   _____

2. What was the source of the bacteria and yeast originally used in dairy product fermentations and breads? _____

   _____

3. What could cause an inferior product in a microbial fermentation process? _____

_____

4. How are microbial fermentations used to preserve foods? _____

_____

Microbes in Soil: The Nitrogen Cycle

Soil is the placenta of life.

PETER FARB

Objectives

After completing this exercise you should be able to:

1. Diagram the nitrogen cycle, showing the chemical changes that occur at each step.
2. Explain the importance of the nitrogen cycle.
3. Differentiate between symbiotic and nonsymbiotic nitrogen fixation.

Background

One aspect of soil microbiology that has been studied extensively is the nitrogen cycle. All organisms need nitrogen for the synthesis of proteins, nucleic acids, and other nitrogen-containing compounds. The recycling of nitrogen by different organisms is called the **nitrogen cycle** (Figure 55.1). Microbes play a fundamental, irreplaceable role in the nitrogen cycle by participating in many different metabolic reactions involving nitrogen-containing compounds. When plants, animals, and microorganisms die, microbes decompose them by proteolysis and ammonification.

Proteolysis is the hydrolysis of proteins to form amino acids (Exercise 15). **Ammonification** liberates ammonia by deamination of amino acids (Exercise 16) or catabolism of urea to ammonia (Exercise 15). In most soil, ammonia dissolves in water to form ammonium ions:

$$NH_3 \;+\; H_2O \;\rightarrow\; NH_4OH \longrightarrow NH_4^+ \;+\; OH^-$$

| Ammonia | Water | Ammonium hydroxide | Ammonium ion | Hydroxyl ion |

Some of the ammonium ions are used directly by plants and bacteria for the synthesis of amino acids.

The next sequence of the nitrogen cycle is the oxidation of ammonium ions in **nitrification.** Two genera of soil bacteria are capable of oxidizing ammonium ion in the two successive stages shown as follows:

$$\overset{\textit{Nitrosomonas}}{2NH_4^+ \;+\; 3O_2 \longrightarrow 2NO_2^-}$$

| Ammonium ion | Oxygen | Nitrite ion |

$$\overset{\textit{Nitrobacter}}{2NO_2^- \;+\; O_2 \longrightarrow 2NO_3^-}$$

| Nitrite ions | Oxygen | Nitrate ions |

These reactions are used to generate energy (ATP) for the cells. Nitrifying bacteria are chemoautotrophs, and many are inhibited by organic matter. Nitrates are an important source of nitrogen for plants.

Denitrifying bacteria reduce nitrates and remove them from the nitrogen cycle. **Denitrification** is the reduction of nitrates to nitrites and nitrogen gas (Exercise 17). This conversion may be represented as follows:

$$NO_3^- \;\rightarrow\; NO_2^- \;\rightarrow\; N_2O \;\rightarrow\; N_2$$

| Nitrate ion | Nitrite ion | Nitrous oxide | Nitrogen gas |

Denitrification is also called **anaerobic respiration.** Many genera of bacteria, including *Pseudomonas* and *Bacillus* are capable of denitrification under anaerobic conditions (Exercises 17 and 19).

Atmospheric nitrogen can be returned to the soil by the conversion of nitrogen gas into ammonia, a process called **nitrogen fixation.** Cells possessing the nitrogenase enzyme can fix nitrogen under anaerobic conditions as follows:

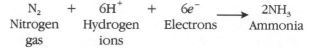

$$N_2 \;+\; 6H^+ \;+\; 6e^- \longrightarrow 2NH_3$$

| Nitrogen gas | Hydrogen ions | Electrons | Ammonia |

Some free-living procaryotic organisms, such as *Azotobacter*, clostridia, and cyanobacteria, can fix nitrogen. Many of the nitrogen-fixing bacteria live in close association with the roots of grasses in the **rhizosphere,** where root hairs contact the soil.

Symbiotic bacteria serve a more important role in nitrogen fixation. One such symbiotic relationship is a mutualistic relationship between *Rhizobium* and the roots of legumes (such as soybeans, beans, peas, alfalfa, and clover) (Figure 55.2). There are thousands of legumes. Farmers grow soybeans and alfalfa to replace nitrogen in their fields. Many wild legumes are able to grow in the poor soils found in tropical rain forests or arid deserts. *Rhizobium* species are specific for the host

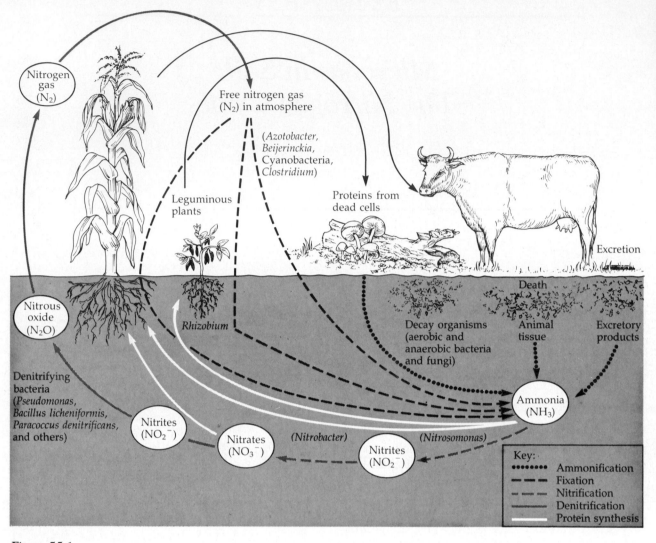

Figure 55.1
The nitrogen cycle.

legume that they infect. When a root hair and rhizobia make contact in the soil, a root nodule forms on the plant. The nodule provides the anaerobic environment necessary for nitrogen fixation.

Symbiotic nitrogen fixation also occurs in the roots of nonleguminous plants. The actinomycete *Frankia* forms root nodules in alders.

Any break in the nitrogen cycle would be critical to the survival of all life. We will investigate the steps of the nitrogen cycle in this exercise.

Materials

Ammonification
Tube containing peptone broth

Nessler's reagent

Ammonium hydroxide

Spot plate

Soil (Bring your own.)

Denitrification
Tubes containing nitrate-salts broth (2)

Nitrate reagents A and B

Soil (Bring your own.)

Nitrogen Fixation
Petri plate containing mannitol–yeast-extract agar

Methylene blue

Sterile razor blade

Legumes (Bring your own; see Figure 55.2.)

Culture

Pseudomonas aeruginosa

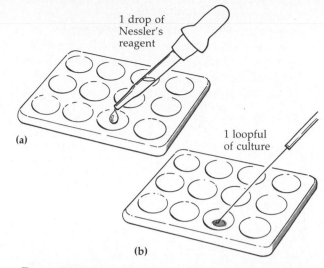

Figure 55.3
(a) Place a drop of Nessler's reagent in a well of a spot plate.
(b) Add a loopful of the sample to be tested, and mix.
Observe for a color change.

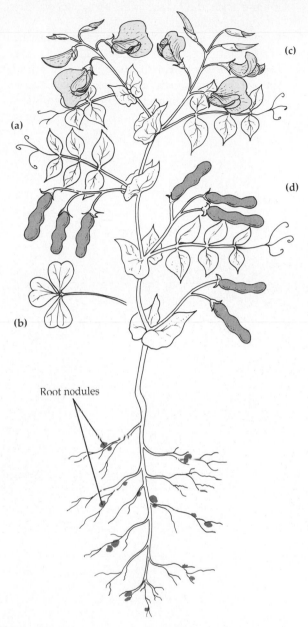

Figure 55.2
Some characteristic features of legumes. The leaves are opposite and may be **(a)** pinnately or **(b)** palmately compound. The flowers **(c)** have five asymmetrical petals, and the fruits **(d)** are pealike.

Demonstrations

Rhizobium inoculum used by farmers

Slides of stained root nodules

Techniques Required

Exercises 3, 10, 11, and 17.

Procedure

Ammonification

1. Obtain an inoculum of soil on a wetted loop and inoculate the peptone broth. What is peptone? _____
   _____

2. Incubate the peptone broth at room temperature and test for ammonia at 2 days and at 7 days.

3. Test for ammonia by placing a drop of Nessler's reagent in a spot plate. Add a loopful of peptone broth, and mix (Figure 55.3). A yellow to brown color indicates the presence of ammonia. Compare your results to a spot to which a drop of ammonium hydroxide and Nessler's reagent have been added. Use an uninoculated tube of peptone broth as a control. Why is a control necessary? _____
   _____

Denitrification

1. Inoculate one nitrate-salts broth with soil using a wetted loop to pick up the soil. Inoculate the other tube with *P. aeruginosa*.

2. Incubate both tubes at room temperature for 1 week.

3. Test for nitrate reduction. Remember how? Describe. _____
   _____

(*Hint:* Refer to Exercise 17.) Record your results.

Nitrogen Fixation

1. Cut off a nodule from a legume and wash it under tap water. (See color plate VI.3.) Observe.

2. Cut the nodule in half with a sterile razor blade. Observe. Crush the nodule between two slides and make a smear by rotating the slides together.

3. Streak a loopful of the crushed nodule on the mannitol–yeast-extract agar. Incubate the plate at room temperature for 7 days.

4. Air dry the slide and heat fix. Stain the slide for 1 minute with methylene blue. Rinse and observe under oil immersion.

5. Observe the demonstrations. What is in the farmer's inoculum? _____

6. Observe the growth on the plate and make a simple stain of the growth with methylene blue. Compare this stain to the stain prepared from the nodule.

Microbes in Soil: The Nitrogen Cycle

Name _____

Date _____

Lab Section _____

Purpose _____

_____

_____

Data

Ammonification

Incubation	Growth	Nessler's Reagent			Ammonia Present?
		Color of Inoculated Peptone	Color of NH_4OH	Color of Control	
2 days					
7 days					

Denitrification

How did you test for denitrification? _____

Inoculum	Growth	Gas	Nitrate Reduction?	Denitrification?
Soil				
Pseudomonas				

Nitrogen Fixation

Diagram of root nodule

Diagram microscopic appearance of
cells in root nodule

Cell morphology _____

Diagram microscopic appearance
of bacteria from culture

Demonstration slides
Label root, nodule, and bacteria

100×

1000×

Cell morphology _____

Describe colonies on the mannitol–yeast-extract plate. _____

_____

_____

Questions

1. Write the equation for ammonification of any amino acid.

2. Why can't you smell ammonia in the peptone broth tubes? _____

_____

_____

3. Why is denitrification a problem for farmers? _____

_____

_____

4. What gas is in the gas tube of a positive denitrification test? _____

_____

_____

5. What genetic engineering project would you propose concerning nitrogen fixation? _____

_____

_____

_____

6. Why do the bacteria in the root nodule look different from the bacteria cultured from the nodule? _____

_____

_____

7. Which step(s) in the nitrogen cycle require oxygen? _____

_____

_____

8. Why is the nitrogen cycle important to all forms of life? _____

_____

_____

Microbes in Soil: Bioremediation

Objectives

After completing this exercise you should be able to:

1. Define bioremediation.
2. Demonstrate enrichment of oil-degrading bacteria.
3. Perform chromatography.

Background

Although many bacteria have dietary requirements similar to ours — which is why they cause food spoilage — others metabolize substances that are toxic to many plants and animals: heavy metals, sulfur, petroleum, and even PCBs and mercury. The use of bacteria to eliminate pollutants is called **bioremediation.** Unlike some forms of environmental cleanup, in which dangerous substances are removed from one place only to be dumped in another, bacterial cleanup eliminates the toxic substance, and often returns a harmless or useful substance to the environment.

In this exercise, we will investigate bacterial degradation of hydrocarbons. A **hydrocarbon** is a compound that contains only carbon and hydrogen atoms. The carbon chains can be straight, branched, or rings. One of the most promising successes occurred on an Alaskan beach following the 1989 Valdez oil spill. Several naturally occurring bacteria in the genus *Pseudomonas* are able to degrade oil for their carbon and energy requirements. In the presence of air, they remove two carbons at a time, in a process called *beta-oxidation*, from a large petroleum molecule. These bacteria degrade the oil too slowly to be helpful in cleaning up an oil spill. However, scientists hit upon a very simple way to speed them up. They dumped nitrogen and phosphorous plant fertilizers onto the beach (a process called *bioaugmentation*). The number of oil-degrading bacteria increased compared to those on unfertilized control beaches, and the test beach is now free of oil.

In this exercise, we will enrich for hydrocarbon-degrading bacteria (Exercise 12). We will use spirit blue agar to test for the presence of these bacteria. Degradation of the hydrocarbon should result in the opaque emulsified oil becoming small, soluble molecules, such as fatty acids and acetyl groups. The color indicator will fade to white as the opacity changes. We will use **chromatography** to determine whether the bacteria are metabolizing the hydrocarbon substrate. In chromatography, samples to be tested are spotted on paper. The bottom of the paper is then placed in a solvent and the solvent moves up the paper. Chemicals in the sample that are the most soluble in the solvent move up the paper quickly, and migration of the least soluble is very slow. Iodine vapor is used to see spots of organic molecules after migration. Differences in migration are known as **R_f values,** which are calculated by dividing the distance a compound migrated by the distance the solvent moved.

Materials

Flask containing minimal salts broth and a hydrocarbon substrate (motor oil or cooking oil)

Petri plates containing spirit blue agar and the same hydrocarbon as the flask (2)

Soil sample, from an area exposed to oil pollution (Bring your own, from a service station or roadside.)

Test tube

Sterile 1-ml pipettes (2)

Chromatography solvent (methylene chloride and acetic acid)

Chromatography jar or Coplin jar

Dextrin developing spray

Iodine staining chamber

Chromatography paper or silica-coated plastic

Micropipette (0.5 — 10 μl)

Micropipette tips (2)

Hockey stick

Techniques Required

Exercises 5, 11, and 12; Appendix A.

Procedure

First Period

1. Label a clean, empty test tube "Uninoculated." Place 1 to 2 ml of medium in the tube for use in the second period. Inoculate 25 ml of oil medium with 2 g soil or mud.

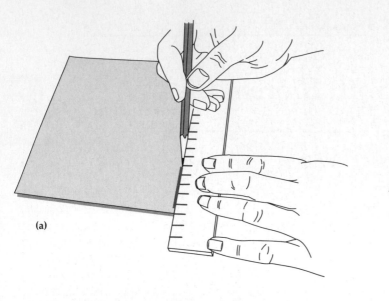

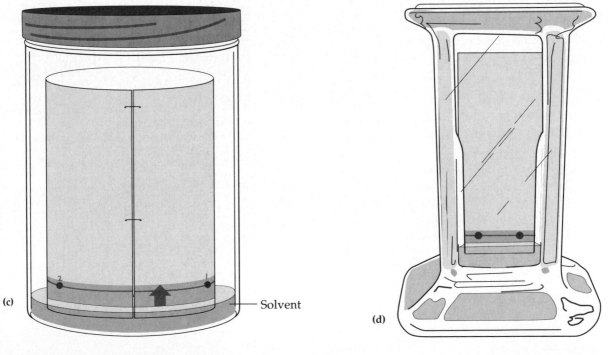

(Solvent)

Figure 56.1

Chromatography. **(a)** Draw a line parallel to the bottom of the paper 1 cm from the bottom. Mark two spots along the line. **(b)** Using a micropipette, apply the sample to one spot. **(c)** Staple the paper so the ends do not touch, and stand the paper in the solvent, or **(d)** stand the silica-coated plastic in a Coplin jar. The level of the solvent should be *below* the sample spots.

2. Inoculate a spirit blue agar plate with 0.5 ml of the inoculated oil medium prepared in step 1.

> ⚠ **Disinfect the hockey stick by dipping in alcohol. Keep the flame away from the beaker.**

Spread the inoculum over the plate using the hockey stick (see Figure 27.2). Incubate at room temperature for 24 hours. (See color plate X.5.)

3. Incubate the flask on a shaker at room temperature for 5 to 7 days.

Second Period

1. Inoculate a spirit blue agar plate with 0.5 ml of the oil medium from the flask on the shaker. Spread the inoculum over the plate using a sterile hockey

stick. Incubate at room temperature for 24 hours.
2. Record the results of the spirit blue agar inoculated during the first period. Gram stain some colonies.

Chromatography (Figure 56.1)
1. Pour chromatography solvent into a jar. The solvent should be about 5 mm deep. Cover the jar.

> ⚠ **Methylene chloride is harmful if inhaled.**

2. Obtain a piece of chromatography paper; touch it only at the edges. Why? _____

With a pencil, draw a line across the paper 1 cm from the bottom of the paper. Label two spots along the line: One for uninoculated medium and one for the medium after incubation (Figure 56.1*a*).
3. Apply 1 μl of uninoculated medium to one spot. Let the spot dry and apply another 1 μl (Figure 56.1*b*).
4. Using a different tip, apply the inoculated medium to the paper using the same procedure (step 3). Let the paper dry.
5. Follow procedure a or b.
   a. If you are using paper, roll the paper into a cylinder and staple the ends so they do not touch each other (Figure 56.1*c*). Place the paper in the tube so that the bottom is in the solvent. The level of the solvent should be *below* the pencil line.

   b. If you are using silica-coated plastic, stand the sheet so that the bottom is in the solvent. The level of the solvent should be *below* the pencil line (Figure 56.1*d*).

6. Let the solvent rise up the paper. Let the solvent rise as far as possible, but *do not let the solvent reach the top of the paper*.
7. Remove the paper and mark the solvent front with pencil. Let the paper dry. You can use a heat lamp to speed up the drying time.
8. Develop the chromatogram by spraying with dextrin; then place the paper in the beaker of iodine.

> ⚠ **Iodine vapor is harmful if inhaled. Keep the jar capped except when placing and removing paper.**

9. When the spots have developed, remove the paper and circle the spots.
10. Measure the distance each spot traveled from the line to the center of the spot. Record your results and calculate the R_f value for each spot.

Third Period
1. Observe and describe the growth on the spirit blue agar plate.
2. Gram stain some colonies.
3. Compare the results with the first spirit blue agar plate.

EXERCISE 56

Microbes in Soil: Bioremediation

Name _____

Date _____

Lab Section _____

Purpose _____

_____

_____

Data

Enrichment

Carbon source used: _____

	Appearance of Oil Medium
Uninoculated	
After incubation for _____ days	

Distance solvent moved: _____ cm

	Distance Moved (cm)	R_f
Uninoculated		
After incubation for _____ days		

Attach your chromatogram.

Spirit Blue Agar

Before enrichment

Colony Description	Oil Degrader?	Morphology and Gram Reaction	Number of This Type

After enrichment

Colony Description	Oil Degrader?	Morphology and Gram Reaction	Number of This Type

Conclusions

Summarize the results of your enrichment. _____

_____

_____

_____

_____

Questions

1. How can you tell which organisms degraded the carbon source on spirit blue agar? _____

_____

_____

2. Write the chemical reaction for hydrolysis of the following lipid:

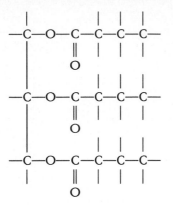

3. Show beta-oxidation of the following hydrocarbon:

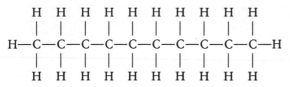

4. Here are the partial formulas of two detergents that have been manufactured:

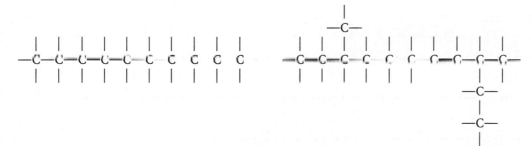

Which of these would be readily degraded by bacteria? Why? _____

_____

_____

5. How could the oil-degrading bacteria you isolated be used to clean up an oil spill? _____

_____

_____

_____

_____

6. Design an experiment to isolate mercury-utilizing bacteria from soil. _____

_____

_____

_____

Appendices

APPENDIX A

Pipetting

Pipettes are used for measuring small volumes of fluid. They are usually coded according to the total volume and graduation units (Figure A.1). To fill a pipette, use a bulb or other mechanical device, as shown in Figure A.2.

☣ **Never use your mouth to fill a pipette.**

Draw the desired amount of fluid into the pipette. Read the volume at the bottom of the meniscus (Figure A.3).

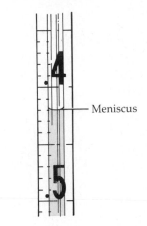

Meniscus

Figure A.3
Read the fluid volume at the lowest level of the meniscus.

Figure A.1
This pipette holds a total volume of 1 milliliter when filled to the zero mark. It is graduated in 0.01-ml units.

(a)

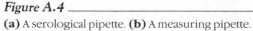

(b)

Figure A.4
(a) A serological pipette. (b) A measuring pipette.

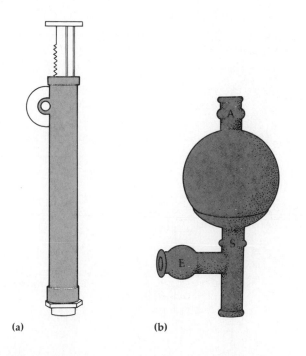

(a)   (b)

Figure A.2
Two types of pipette aspirators. (a) Attach this plastic pump to the pipette, and turn the wheel to draw fluid into the pipette. Turning the wheel in the other direction will release the fluid. (b) Insert a pipette into this bulb. While pressing the A valve, squeeze the bulb, and it will remain collapsed. To draw fluid into the pipette, press the S valve; to release fluid, press the E valve.

Fill the pipette to above the zero mark, then allow it to drain just to the zero. The desired amount can then be dispensed.

Microbiologists use two types of pipettes. The **serological pipette** is meant to be emptied to deliver the total volume (Figure A.4*a*). Note that the graduations stop above the tip. The **measuring pipette** delivers the volume read on the graduations (Figure A.4*b*). This pipette is not emptied, but the flow must be stopped when the meniscus reaches the desired level.

Aseptic use of a pipette is often required in microbiology. Bring the entire closed pipette container to your work area. If the pipettes are wrapped in paper, open the wrapper at the end opposite the delivery end; in a canister, the delivery end will be at the bottom of the canister. Do not touch the delivery end of a sterile pipette. Remove a pipette and, with your other hand, pick up the sample to be pipetted. Remove the cap from the sample with the little finger of the hand holding the pipette. Fill the pipette and replace the cap on the sample.

> ☣ **After pipetting, place the contaminated pipette in the appropriate container of disinfectant. If it is a disposable pipette, discard it in a biohazard bag or any container designated for biohazards.**

Micropipettes are used to measure volumes less than 0.5 ml. The design and operation of micropipettes varies according to the manufacturer. Most use a disposable tip to hold the fluid, as shown in Figure A.5. To use a micropipette, select one that holds the volume you need. The micropipette is labeled with its range, for example, 0.5–10 μl, 2–10 μl, 10–100 μl, 100–1,000 μl. Set the desired volume on the digital display (Figure A.6*a*) by turning the control knob (Figure A.6*b*).

Questions

1. Which micropipette would you use to measure 15 μl? _____

2. If this is the display for the micropipette you chose for question 1, write in the 1 and 5 to show 15 μl.

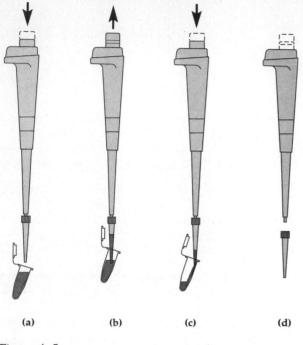

(a) (b) (c) (d)

Figure A.5 _____
Generalized use of a micropipette. **(a)** Depress the control button and insert the tip into the liquid. **(b)** Smoothly release the button to allow the liquid to enter the tip. **(c)** Place the tip against the inside of the receiving tube and depress the button. **(d)** Eject the used tip by pressing the eject button or by pressing the control button to the final stop.

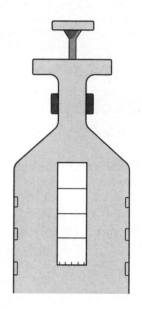

3. How do you eject the tip on your micropipette? _____

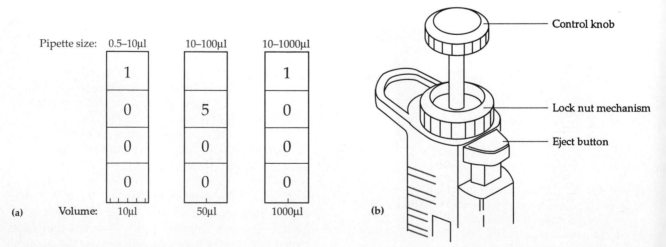

Pipette size:	0.5–10μl	10–100μl	10–1000μl
	1		1
	0	5	0
	0	0	0
	0	0	0
(a) Volume:	10μl	50μl	1000μl

Control knob

Lock nut mechanism

Eject button

(b)

Figure A.6 _____
Adjusting a micropipette. **(a)** Reading the display. **(b)** Turning the control knob on this micropipette changes the volume picked up when the control knob is depressed.

APPENDIX B

Dilution Techniques and Calculations

Bacteria, under good growing conditions, will multiply into such large populations that it is often necessary to dilute them to isolate single colonies or to obtain estimates of their numbers.* This requires mixing a small, accurately measured sample with a large volume of sterile water or saline called the **diluent,** or **dilution blank.** Accurate dilutions of a sample are obtained through the use of pipettes. For convenience, dilutions are usually made in multiples of ten.

A single dilution is calculated as follows:

$$\text{Dilution} = \frac{\text{Volume of the sample}}{\text{Total volume of the sample and the diluent}}$$

For example, the dilution of 1 milliliter into 9 milliliters equals

$$\frac{1}{1 + 9}, \text{ which is } \frac{1}{10} \text{ and is written 1:10}$$

Experience has shown that better accuracy is obtained with very large dilutions if the total dilution is made out of a series of smaller dilutions rather than one large dilution. This series is called a **serial dilution,** and the total dilution is the product of each dilution in the series. For example, if 1 ml is diluted with

*Adapted from C. W. Brady. "Dilutions and Dilution Calculations." Unpublished paper. Whitewater, WI: University of Wisconsin, n.d.

9 ml, and then 1 ml of that dilution is put into a second 9-ml diluent, the final dilution will be

$$\frac{1}{10} \times \frac{1}{10} = \frac{1}{100} \quad \text{or} \quad 1:100$$

To facilitate calculations, the dilution is written in exponential notation. In the example above, the final dilution 1:100 would be written 10^{-2}. (Refer to the box, "Exponents, Exponential Notation, and Logarithms.") A serial dilution is illustrated in Figure B.1.

Procedure

1. Aseptically pipette 1 ml of sample into a dilution blank.
   a. If the dilution is into a tube, mix the contents by rolling the tube back and forth between your hands (Figure 11.4).
   b. If the dilution is into a 99-ml blank, hold the cap in place with your index finger and shake the bottle up and down through a 35-cm arc (Figure 53.2).
2. It is necessary to use a fresh pipette for each dilution in a series, but it is permissible to use the same pipette to remove several samples from the same bottle, as when plating out samples from a series of dilutions.

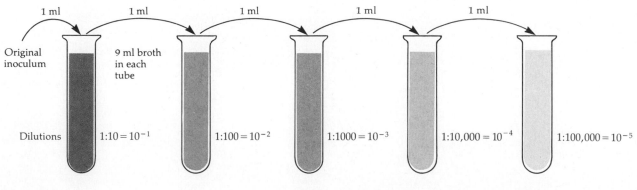

Figure B.1

A 1-ml sample from the first tube will contain ¹⁄₁₀ the number of cells present in 1 ml of the original sample. A 1-ml sample from the last tube will contain 1/100,000 the number of cells present in 1 ml of the original sample.

351

Problems

Practice calculating serial dilutions using the following
problems.

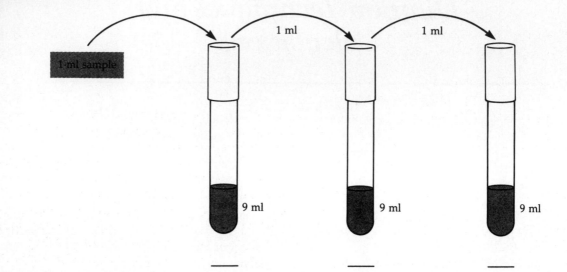

1.
___   ___   ___

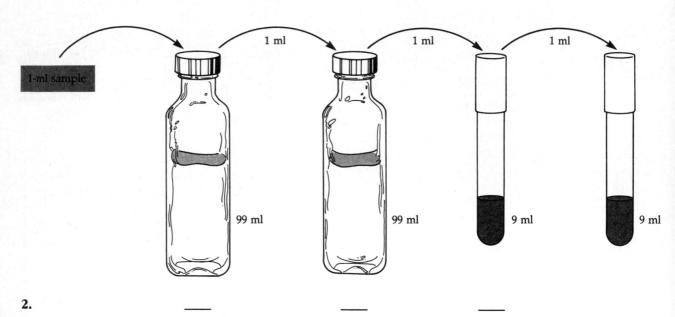

2.
___   ___   ___

3. Design a serial dilution to achieve a final dilution
of 10^{-8}.

Exponents, Exponential Notation, and Logarithms

Very large and very small numbers—such as 4,650,000,000 and 0.00000032—are cumbersome to work with. It is more convenient to express such numbers in exponential notation, that is, as a power of 10. For example, 4.65×10^9 is written in **standard exponential notation,** or **scientific notation;** 4.65 is the **coefficient,** and 9 is the power, or **exponent.** In standard exponential notation, the coefficient is a number between 1 and 10, and the exponent is a positive or negative number.

To change a number into exponential notation, follow two steps. First, determine the coefficient by moving the decimal point so you leave only one nonzero digit to the left of it. For example:

$$0.00000032$$

The coefficient is 3.2. Second, determine the exponent by counting the number of places you moved the decimal point. If you moved it to the left, the exponent is a positive number. If you moved it to the right, the exponent is negative. In the example, you moved the decimal point 7 places to the right, so the exponent is -7. Thus

$$0.00000032 = 3.2 \times 10^{-7}$$

Now suppose we are working with a very large number instead of a very small number. The same rules apply, but our exponential value will be positive rather than negative. For example:

$$4,650,000,000. = 4.65 \times 10^{+9}$$
$$= 4.65 \times 10^9$$

To multiply numbers written in exponential notation, multiply the coefficients and *add* the exponents. For example:

$$(3 \times 10^4) \times (2 \times 10^3) =$$
$$(3 \times 2) \times 10^{4+3} = 6 \times 10^7$$

To divide, divide the coefficients and *subtract* the exponents. For example:

$$\frac{3 \times 10^4}{2 \times 10^3} = \frac{3}{2} \times 10^{4-3} = 1.5 \times 10^1$$

Microbiologists use exponential notation in many kinds of situations. For instance, exponential notation is used to describe the number of microorganisms in a population. Such numbers are often very large. Another application of exponential notation is to express concentrations of chemicals in a solution—chemicals such as media components, disinfectants, or antibiotics. Such numbers are often very small. Converting from one unit of measurement to another in the metric system requires multiplying or dividing by a power of 10, which is easiest to carry out in exponential notation.

A **logarithm** is the power to which a base number is raised to produce a given number. Usually we work with logarithms to the base 10, abbreviated **log₁₀.** The first step in finding the $\log_{10}$ of a number is to write the number in standard exponential notation. If the coefficient is exactly 1, the $\log_{10}$ is simply equal to the exponent. For example:

$$\log_{10} 0.00001 = \log_{10}(1 \times 10^{-5})$$
$$= -5$$

If the coefficient is not 1, as is often the case, a logarithm table or calculator must be used to determine the logarithm.

Microbiologists use logs for pH calculations and for graphing the growth of microbial populations in culture.

Source: G. J. Tortora, B. R. Funke, and C. L. Case. *Microbiology: An Introduction,* 4th ed. Redwood City, CA: Benjamin/Cummings, 1992, Appendix F.

Use of the Spectrophotometer

In a **spectrophotometer,** a beam of light is transmitted through a bacterial suspension to a photoelectric cell (Figure C.1). As bacterial numbers increase, the broth becomes more turbid, causing the light to scatter and allowing less light to reach the photoelectric cell. The change in light is registered on the instrument as **percentage of transmission,** or **%T** (the amount of light getting through the suspension) and **absorbance** (a value derived from the percentage of transmission). Absorbance is a logarithmic value and is used to plot bacterial growth on a graph.

Operation of the Spectronic 20 and Meter-Model Spectronic 21 Spectrophotometers (Figure C.2)

1. Turn on the power and allow the instrument to warm up for 15 minutes.
2. Set the wavelength for maximum absorption of the bacteria and minimal absorption of the culture medium.

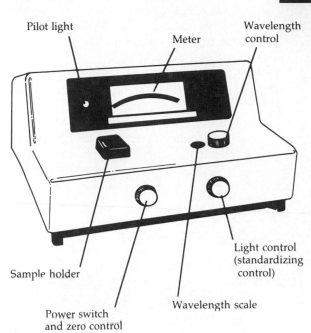

Figure C.2

Bausch and Lomb Spectronic 20.

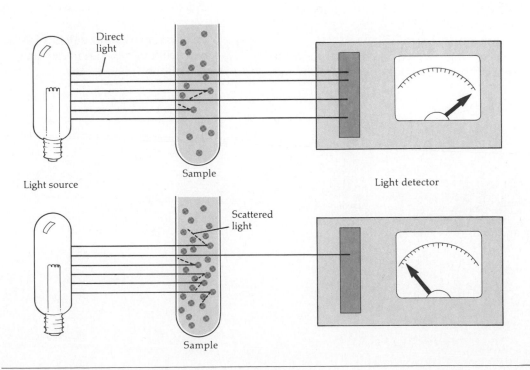

Figure C.1

Estimation of bacterial numbers by turbidity. The amount of light picked up by the light detector is proportional to the number of bacteria. The less light transmitted, the more bacteria in the sample.

3. For the Spectronic 20, *zero* the instrument by turning the zero control until the needle measures 0% transmission. To read the scale, have the needle at eye level, and look at it so that the needle is directly over its reflection in the mirror.
4. To read a metered scale, look directly at the meter so the needle is superimposed on its reflection in the mirror.
5. Place an uninoculated tube of culture medium (*control*) in the sample holder. To *standardize* the instrument, turn the light control until the needle registers 100% transmission.
6. To take a sample measurement, place an inoculated tube of culture medium in the sample holder and record the %T from the scale.
7. %T is usually read on the Spectronic 20 because it is the more accurate scale. Calculate the absorbance as follows:

$$\text{Absorbance} = -\log\left(\frac{\%T}{100}\right)$$

Operation of the Digital-Model Spectronic 21 (Figure C.3)

1. Turn on the power and allow the instrument to warm up for 15 minutes.
2. Set the wavelength for maximum absorption of the bacteria and minimal absorption of the culture medium.
3. Select the operating mode: A (absorbance) or %T (transmission). Absorbance data are more linear and usually preferable.
4. Set the sensitivity switch to LO.
5. Place an uninoculated tube of culture medium (*control*) in the sample holder. To *standardize* the

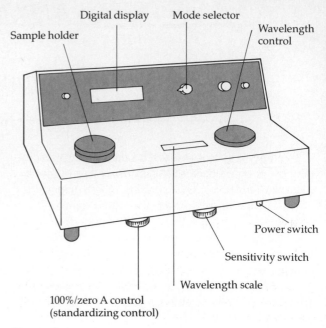

instrument, turn the 100% T/Zero A control until the display registers 000 A. If 000 A can't be reached, turn the sensitivity to M or HI. A U in the upper-left corner of the display indicates that you are under the range of the light detector, and an O in the display indicates that you are over the range. In either case, use the sensitivity switch and the standardizing control to make the necessary adjustments.
6. To take a sample measurement, place an inoculated tube of culture medium in the sample holder and record the absorbance (Abs.) from the display.

Figure C.3 _____

Bausch and Lomb Spectronic 21, digital model.

APPENDIX D

Graphing

A **graph** is a visual representation of the relationship between two variables. Whenever one variable changes in a definite way in relation to another variable, this relationship may be graphed.

Microbiologists work with large populations of bacteria and frequently use graphs to illustrate the activity of these populations. Graphs dealing with populations of cells are drawn on semi-log graph paper. The horizontal, or **X-axis,** is a linear scale. The X-axis is used for the **independent variable,** that is, the variable not being tested. The vertical, or **Y-axis,** is a logarithmic scale where each cycle represents a power of 10. The dependent variable is marked off along the Y-axis. The **dependent variable** changes in relation to the independent variable. The intersection point of the X-axis and Y-axis is called the **origin,** and all units are marked off from this point.

To draw a graph, first establish the two variables you want to graph — for example, time and number of bacteria (Table D.1). The independent variable is time, and the dependent variable is the number of bacteria. Mark the appropriate units of measure along the axes (Figure D.1).

We will graph the numbers of bacteria in the culture incubated at 20°C. The first point on the graph is (0, 5.00×10^5). Locate 0 on the X-axis, measure directly above it the distance 5.00×10^5, and mark this point. After a sufficient number of points have been plotted, a smooth curve can be drawn connecting these points. The line through the points should be straight and does

Table D.1

Numbers of *E. coli* in Three Cultures Incubated at Different Temperatures

Time (hours)	Temperature (°C)		
	20	15	35
0	5.00×10^5	1.50×10^5	5.40×10^5
4	—	1.50×10^5	—
5	4.60×10^5	—	—
6	—	—	5.20×10^5
10	5.78×10^5	5.62×10^5	6.00×10^5
11	8.80×10^5	—	1.23×10^6
12	1.05×10^6	—	2.58×10^6
13	1.16×10^6	—	4.93×10^6
14	2.32×10^6	—	9.00×10^6
15	3.80×10^6	1.62×10^6	—
16	7.00×10^6	—	—
17	8.10×10^6	—	—
18	9.60×10^6	—	—
19	1.95×10^7	—	—
20	—	2.70×10^6	—

not have to connect all the points. When the points do not fall in a straight line, draw a line between the points, leaving an equal number of points above and below the line.

When graphs are drawn that compare the same variables under different conditions (for example, bacterial growth rates at different temperatures), it is best to draw all the graphs on the same paper so comparisons are obvious and easily made. Graphs should be titled to explain the data presented.

The data for cultures incubated at 15°C and 35°C (Table D.1) can be plotted on the graph in Figure D.1. It is not necessary to have bacterial numbers for the same

"time points" in each graph. The missing points will be filled in by the line. This is possible because the line between each point is an interpretation of what was happening between measurements.

Activity

Using the data provided in Table D.1, graph the growth curves on the graph paper and compare the effect of temperature on the rate of growth. Since *rate* is a change over time, the rate can be interpreted by the slope of the line. What will you title this graph? _____

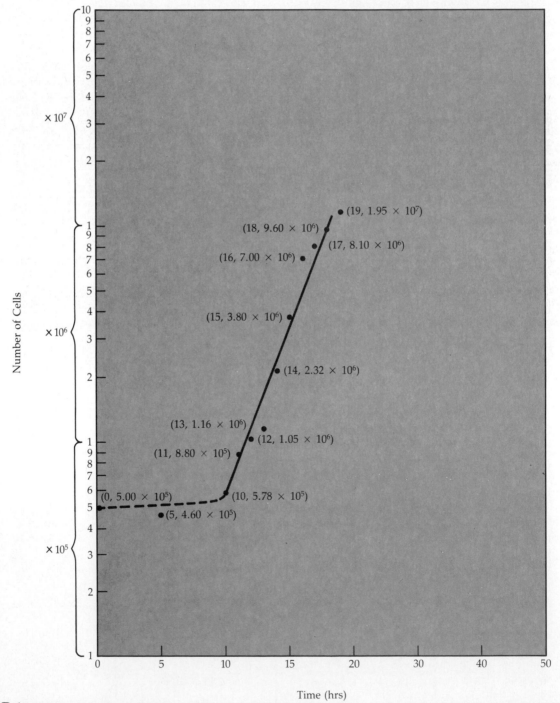

Figure D.1 _____

Graph showing growth of *E. coli* at 20°C.

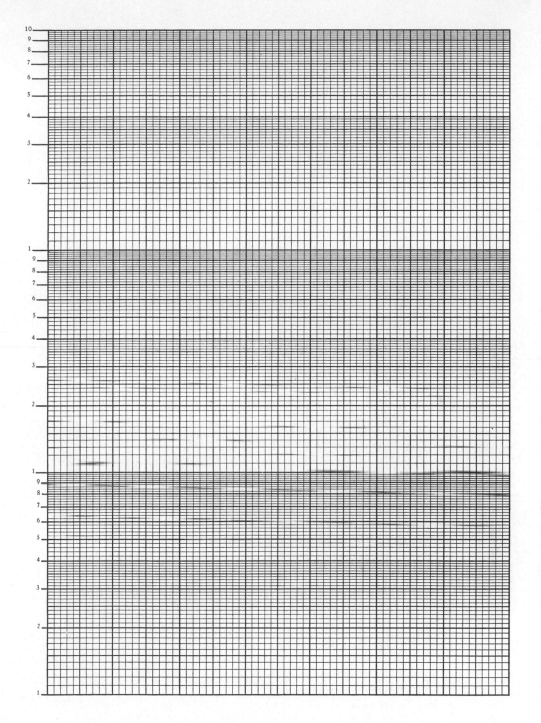

APPENDIX E

Use of the Dissecting Microscope

The **dissecting microscope** (Figure E.1) is useful for observing specimens that are too large for the compound microscope and too small to be seen with the naked eye. The dissecting microscope is also called a **stereoscopic** microscope because the image is three-dimensional.

Dissecting microscopes are usually equipped with 10× ocular lenses and a 1× objective lens. Some dissecting microscopes have a rotating nosepiece that houses 1× and 2× objective lenses.

The dissecting microscope may have a built-in light source or a substage mirror and auxiliary lamp. Illumination can be adjusted on microscopes with two built-in lamps and/or by using a mirror. For transparent specimens, the light should be directed through the specimen from under the stage. For opaque specimens, light should be directed onto the specimen from the top. Both types of lighting can be used for observing a translucent specimen.

The specimen is placed on the stage. Some microscopes have an alternate black stage plate for use with opaque specimens. The stage clips can be moved to the side when large specimens, such as Petri plates, are used. While looking through the microscope, adjust the width of the eyepieces so you see one field of vision. If two circles or fields are visible, the ocular lenses are too far apart; if the two fields overlap, the ocular lenses are too close together.

To focus, turn the adjustment knob.

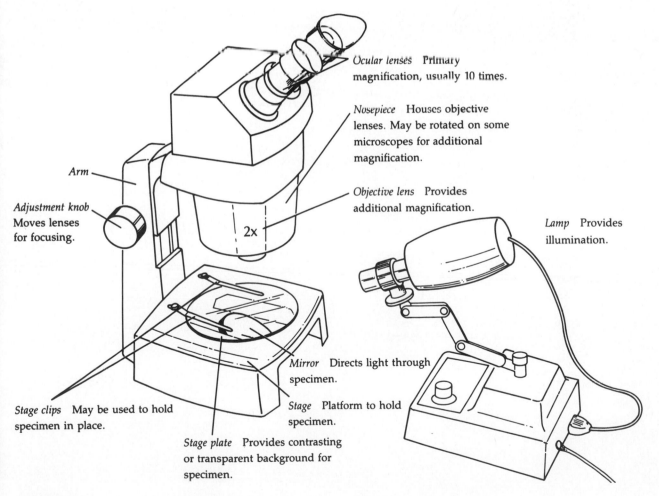

Ocular lenses Primary magnification, usually 10 times.

Nosepiece Houses objective lenses. May be rotated on some microscopes for additional magnification.

Objective lens Provides additional magnification.

Lamp Provides illumination.

Arm

Adjustment knob Moves lenses for focusing.

2x

Mirror Directs light through specimen.

Stage clips May be used to hold specimen in place.

Stage Platform to hold specimen.

Stage plate Provides contrasting or transparent background for specimen.

Figure E.1
The dissecting microscope: principal parts and their functions.

Questions

1. What is the magnification of your dissecting microscope? _____

2. Using a ruler, measure the size of the field of vision at low power: _____ mm. At high power: _____ mm.

3. How can you adjust the direction of illumination on your microscope? _____
_____

APPENDIX F

Use of the Membrane Filter

APPENDIX

F

Membrane filtration can be used to separate bacteria, algae, yeasts, and molds from solutions. Cellulose or polyvinyl membrane filters with 0.45 μm pores are commonly used to trap microorganisms. In Exercise 37, membrane filtration is used to separate viruses from their host cells because the viruses will pass through the filter. In Exercise 52, membrane filters are used to trap bacteria found in water in order to count the bacteria.

Membrane filters have many uses in industrial microbiology. They are used to filter wine, soft drinks, air, and water to detect microorganisms and particulate matter. In addition, membrane filters can be used to separate large molecules, such as DNA, from solutions.

The receiving flask, filter base, and filter support can be wrapped in paper and sterilized by autoclaving (Figure F.1). Membrane filters can also be sterilized by autoclaving. For use, the filter base must be unwrapped aseptically. With sterile, flat forceps, a sterile filter is placed on the filter support, and the filter funnel is clamped or screwed into place.

Gravity alone will not pull a sample through the small filter pores, so the filtering flask is attached to a vacuum source (Figure F.1). Since air pressure is then greater outside the flask, the sample is pushed into the vacuum inside the flask. A vacuum pump or aspirator connected to a water faucet is usually used to provide the necessary vacuum. An aspirator trap is placed between the vacuum source and the receiving flask to prevent the flow of solution into the pump or sink. With small filters and small volumes of fluid, a syringe can be used to supply the needed vacuum.

Following filtration, the filtrate in the flask is free of all microorganisms larger than viruses. For observation of cells trapped on the filter, the filter can be dipped into immersion oil to make it translucent and then mounted on a microscope slide. Microorganisms can be cultured from the filter by placing the filter on a solid nutrient medium (Exercise 52).

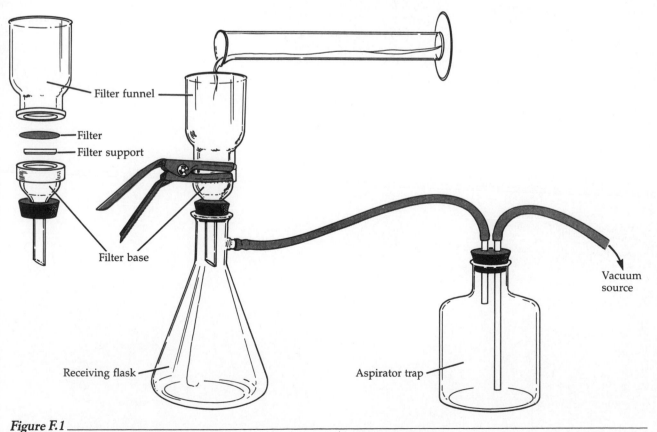

Figure F.1

Membrane filtration setup. (Plastic disposable filters can be substituted for glass filters.)

363

Electrophoresis

Charged macromolecules can be separated according to size and charge by a process called **electrophoresis.** The material to be tested is placed at one end of a gel matrix. Polyacrylamide gel or agarose are used to separate proteins and small pieces of DNA and RNA. Agarose is the usual matrix for separation of nucleic acids. A buffer (salt solution) is used to conduct electric current through the matrix. When an electric current is applied, the molecules start to move. Molecules with different charges and sizes will migrate at different rates.

Procedure

1. Place the well-forming comb in position on a glass slide or a casting tray (Figure G.1). If you are using a casting tray, seal the open ends with masking tape. Pour melted agarose onto the glass plate until the agar surrounds the teeth. The agar should be about 5 mm deep and have a flat surface. Surface tension will hold the agarose on the glass. Let the gel

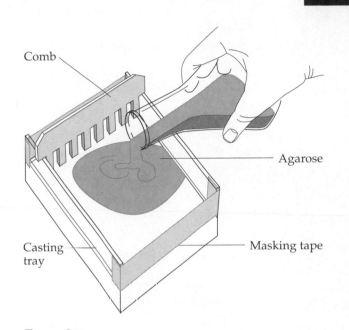

Figure G.1
Preparation of an agarose minigel.

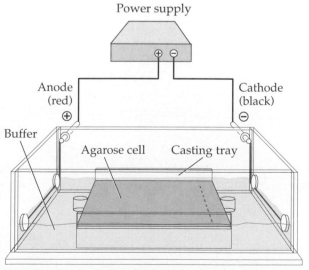

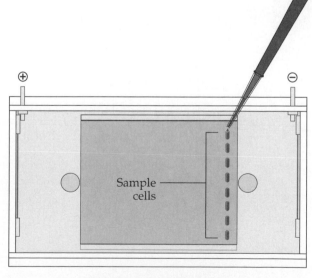

Figure G.2
Setup of electrophoresis equipment.

harden for about 20 minutes. When the gel is solid, carefully remove the comb and the tape.

2. Fill the buffer chamber with the appropriate buffer. Add just enough buffer to cover the gel. Place the agar in the buffer chamber with or without the casting tray.

3. Fill the wells with samples to be tested. Close the lid and apply current (Figure G.2).

4. During electrophoresis, a visible marker dye will migrate in front of the sample. If the dye isn't moving in the right direction, you have accidently reversed the polarity of the electric field. If the marker doesn't move and if there is little current, you have an open circuit or have not added the correct buffer. In this case, you must start over.

5. The two components of the tracking dye migrate at different rates. Why? _____ Turn off the power when the two dyes are approximately 3 cm apart. Remove the gel. The gel can now be stained to locate the macromolecules. Store stained or unstained gels in sealed plastic bags in the refrigerator.

Index